Vedic Mathematics for all Ages

A Beginners' Guide

India's Scientific Heritage

General Editor: Dr L M Singhvi

12

Editorial Panel

Vedic Mathematics for all Ages

A Beginners' Guide

Vandana Singhal

MOTILAL BANARSIDASS PUBLISHER
PRIVATE LIMITED - DELHI

Reprint: Delhi, 2008
First Edition: Delhi, 2007

ISBN: 978-81-208-3211-4 (Cloth)
ISBN: 978-81-208-3230-5 (Paper)

MOTILAL BANARSIDASS

41 U.A. Bungalow Road, Jawahar Nagar, Delhi 110 007
8 Mahalaxmi Chamber, 22 Bhulabhai Desai Road, Mumbai 400 026
236, 9th Main III Block, Jayanagar, Bangalore 560 011
203 Royapettah High Road, Mylapore, Chennai 600 004
Sanas Plaza, 1302 Baji Rao Road, Pune 411 002
8 Camac Street, Kolkata 700 017
Ashok Rajpath, Patna 800 004
Chowk, Varanasi 221 001

Printed in India
BY JAINENDRA PRAKASH JAIN AT SHRI JAINENDRA PRESS,
A-45 NARAINA, PHASE-I, NEW DELHI 110 028
AND PUBLISHED BY NARENDRA PRAKASH JAIN FOR
MOTILAL BANARSIDASS PUBLISHERS PRIVATE LIMITED,
BUNGALOW ROAD, DELHI 110 007

<u>Dedication</u>

*I would like to dedicate this book
to all things wise and beautiful just like
my father Late Shri Y.K.Garg
and my father- in- law
Late Shri Shashikant Singhal
who always believed in me
and encouraged me to do my best.*

Foreword

Mathematics, like all other sciences, developed through a staggered progression. Different civilizations, at different points in time, had their own approaches to the basic mathematical structures and it is only in the last two centuries or so that a somewhat codified approach to mathematics education has evolved in the world. Needless to say each civilization developed techniques to compute mathematical results in their own way. Some of them are elegant and beautiful and some of them could be of dubious value. The ancient Indian civilizations, especially the ones closely linked with Vedic heritage, had their own approach to computation and in recent years there has been a resurrection of interest in this branch of mathematics.

Ms. Vandana Singhal has attempted to codify several useful results embedded in the ancient lore, in a form which is easily accessible to the children learning mathematics. Many of the chapters deal with computations using simple techniques which will shorten the effort involved in the conventional approach. The price one pays, of course, is that one has to learn the tricks, memorize them and use the appropriate one for each problem. While one might think that this takes away the generality of the modern approach, it certainly has the element of charm and intrigue which children [and grown-ups!] will find entertaining. Even working out why many of these approaches lead to correct results is a valuable exercise by itself.

Ms. Singhal has presented these in a set of easy chapters with appropriate inter-relationship and structure. There are also exercises in each of the chapters which I thought would go a long way in keeping the reader amused. In these days, when students hardly memorize multiplication tables and start pressing the calculator buttons for every computation, this book will come as a refreshingly different approach towards enjoying mathematics and computation.

Dr. Thanu Padmanabhan
Dean, Inter University Center
of Astronomy and Astrophysics,
Pune.

Preface

From a very early age I have been fascinated by numbers and the magic around them. Mathematical puzzles have always been of interest to me. Being a mensan I always try to find logical explanations for thing and mathematics being so logical and creative has attracted me the most.

Seeing my interest in mathematics my late father-in-law Shri Shashikant Singhal presented me with a book on Vedic Mathematics. It was really astonishing to read it and to learn various short methods of calculations from it.

In school and college, I had seen my friends fear mathematics as they found it very difficult. I did not want my children to feel the same and it was then that I decided to teach Vedic methods of calculations to them. I could soon see the difference in them. Their competency improved not only in mathematics but also in other subjects.

Thereafter I started teaching these wonderful methods to other children as well, So far I have taught Vedic methods of calculations to more than 1000 students in India and in the U.S.A and observed the development it has made in them. Vedic methods gave them a more logical, creative and innovative approach towards any problem since in Vedic mathematics, we can solve one sum in many different ways.

When I conduct workshops for children they feel the solutions are like magic, but when they know the methods, it becomes *Mathe-magic*.

The joy that I saw in the eyes of children after learning Vedic methods inspired me to write this book. In my attempt to make the book user friendly, I have used different colours which will be visually appealing to children and there is significance to the colours that are used. I have used **bold blue colour** for the answers in each step and as carry over plays a very important part in mathematics, I have used blue colour for carry over.

The reader can also understand the steps by looking at the graphics alone even if he does not read the explanation of each step he can still enjoy the simplicity and beauty of the methods. As children always steer away from practice, I have added creative activities in each chapter to give them a good practice at the end of each chapter. They would not even realize how many sums they have done while doing these colourful activities.

The origami activities will enable the reader to visualize and make different geometric figures which will help them in geometry. However, this is a beginner's guide to Vedic mathematics and concepts on higher topics like Geometry, Trigonometry, Calculus, Conics, Coordinate

Geometry, Partial Fractions, Random Cube Roots and others will be dealt with in my second book.

This book would not have been possible but for the efforts and help rendered by my family and friends.

I would especially like to thank Dr. Thanu Padmanabhan, a distinguished scientist, and Dean of IUCAA (Inter University Center of Astronomy and Astrophysics), for taking time out from his extremely busy schedule and going through the entire book and writing an informative foreword to the book.

A special thanks to Prof. Kenneth Williams, International faculty of the Academy of Vedic Mathematics based in London, U.K., who spared his time to go through the book and gave me valuable suggestions, to make the book more interesting.

A very special thanks to my dear friend Meena Harisinghania, for editing my book, without who's help this book would not have taken its present shape.

I would like to thank Dr. Narayan Desai, Executive Council member, Mensa India (Mensa, an organization about people of high IQ all over the world), with whom I conducted several workshops for mensans.

I thank my friend Mr. Vivek Naidu for encouraging me to write this book and for his exemplary help from its initial stage. A special thanks to Mr. Gopal Ramabadran for all his help and support.

I thank my brother Mr. Vishal Garg and his beautiful wife Shilpa for designing the most appropriate cover page of this book. I thank my mother who always supported me in all that I wanted to do and taught me to always look ahead and not to look back.

This book would not have been possible without the inspiration and support of my husband Sharad and my two sons Shashwat aged 15, and Sarthak aged 11, on whom I conducted all the experiments regarding the methodology presented in this book. I would like to thank them for their valuable suggestions to make it more child friendly. I would also like to thank all those who have been directly or indirectly involved in making this book possible.

Vandana Singhal
Pune, 2006

Feedback...

1. It is now a proven fact that Vedic Mathematics is a key component in nurturing logical and analytical capabilities of Mensans. I strongly recommend this book for developing logical and analytical thinking. This Vedic Mathematics book will be a milestone in spreading Vedic knowledge to a greater part of the society."

    Dr. Narayan Desai,
    Executive Council member, Mensa India

2. The Vedic Mathematics workshop conducted for students of class V to XII in the school was very informative and useful for the students.

    Mrs. Meera Sain,
    Principal, Army Public School, Pune

3. Student's creativity and grasping power increases after they learn Vedic mathematics .....

    23rd April 2006, Maharashtra Herald

4. Students in Pune are fast subscribing to the Vedic system of learning .... mathematics is shedding its scary image , thanks to this age old technique....

    25th October 2004, Pune Times

5. This workshop is quite different from the others, and is very useful for children in the future.

    Mrs. Jammalamadaha, mother of Mana,
    Class VIII student, U.S.A

6. It increased my calculation speed. Now I can do simple calculations without a calculator and faster than it.

    Vinay Asrani, IInd Year, B.Com

7. New Interface with ancient but scientific mathematical techniques. Improved analytical and speed calculation skills.

    Harshad Sabnis, B.Com, MBA

8. I enjoyed the workshop very much and will use the origami activities to decorate my home.

> Kaustubh Divekar, Class VI student,
> Atul Vidyalaya, Valsad, Gujrat

9. I enjoyed the workshop very much. It was a playful method of learning mathematics.

> Parth Chandra, Class VIII student,
> Center Point School, Nagpur

10. I used to dislike mathematics but after attending the workshop I have become extremely interested in mathematics.

> Virendra Naik, Class VII student,
> Bhojwani Academy, Pune

11. My son enjoyed the methods. It is very useful.

> Mrs. Yagnik, mother of Gunjan,
> Class VII student, U.S.A

12. We are now able to do difficult calculations very easily and mentally.

> Anansa Ahmed, Class XII, Pune

13. Very informative and develops interest in mathematics.

> Ankita Gupta, P.E.I.
> (Chartered Accountancy)

14. Very useful as the methods are very easy and help in fast calculations.

> Rahul Shirodkar, Class IX Student,
> Sahyadri School, Rajgurunagar

15. Excellent, simple and fast techniques.

> Sunil Bhat, Engineer,
> TATA MOTORS, Pune

16. Gave some insight for fast calculations and made intelligent guesswork easy.

> Dr Shirish Darak,
> M.A. (Medical Anthropology)

CONTENTS

Chapters

Introduction

Vedic Mathematics is an ancient system of Mathematics. It is a gift to the world and was formulated over many centuries by the ancient sages and rishis of India. It was rediscovered, from the Vedas between 1911 and 1918 by Jagadguru Swami Sri Bharati Krishna Tirthaji Maharaja.

Swamiji was the Shankaracharya of the Govardhan Math, Jaganath Puri as well as Dwarka, Gujrat (1884-1960). He was an exceptionally brilliant student and a great scholar of Sanskrit. At the young age of twenty one he passed M.A. in seven subjects including Science, Mathematics, English, History, and Philosophy simultaneously, securing the highest honors in all. Swamiji who was an accomplished Vedic scholar, wrote 16 volumes on Vedic mathematics comprehensively covering all branches of mathematics. As history goes they were all unfortunately mysteriously lost. Despite his failing health and weak eyesight, Swamiji with his untiring capacity, will, and determination wrote a comprehensive book on Vedic Mathematics covering virtually all the aspects of the lost 16 volumes in one compact volume.

The Vedas are a store house of all knowledge needed by mankind. They are four in number and all the four Vedas namely Rig Veda, Yajur Veda, Sama Veda and Atharva Veda, consist of Samhitas, Brahmanas, Aranyakas, Upvedas and Upanishads. Of these four, the first three namely Samhitas, Brahmanas and Aranyakas contain several thousand Mantras or Hymns, Ritual practices and their interpretations. Vedic mathematics forms the part of the Sthapatyaveda, an Upveda of Atharva Veda.

The term Vedic Mathematics refers to a set of sixteen mathematical formulae or sutras and their corollaries or sub sutras derived from the Vedic system. This speaks for its coherence and simplicity in handling mathematical problems. The sutras not only develop aptitude and ability but also, nurture and develop our logical thinking and intelligence and also encourage innovativeness.

The sutras being single line phrases are easy to understand and remember. Although the sutras are in Sanskrit, the knowledge of Sanskrit language is not a must as they are very well translated. The sutras are so beautifully interrelated and unified, that any mathematical operation can be performed in many ways using different sutras and each sutra can be applied to solve many different mathematical operations.

Vedic Mathematics while nurturing our brain also helps us to relate to our past. A full concentration on it even helps us develop a spiritual bend within ourselves, for example a devotional hymn in praise of Lord Krishna, when decoded gives the value of pi (π) up to 32 decimal places in Trigonometry.

These qualities make mathematics easy, enjoyable and flexible and their qualitative approach makes use of both parts of the brain. So versatile is this science that it has been incorporated in the educational syllabi of many countries worldwide. Even NASA scientists applied its principles in the area of artificial intelligence.

In the Vedic system 'difficult' problems or huge sums can often be solved very quickly and calculations can be carried out mentally or involve one or two steps. Its simplicity leads to more creative, interested and intelligent pupils.

Parents have the misconception that children will get confused by new methods of calculations, but, on the contrary, they are able to correlate the methods learnt at regular school with the Vedic ones. As a result the student develops mathematical intelligence and gets more confidence in the subject. Students trained with the Vedic Mathematics advantage are often ready with the answers soon after the teacher finishes writing the problem on the board.

Vedic Mathematics is useful in preparing students for competitive examinations. It provides them with the "extra something" which helps them to be different. People argue, "Why should I learn Vedic mathematics in this age of calculators?" It may help them to remember that many big digit calculations can be done much faster by Vedic methods than by calculators. Calculators also have a limit to the number of digits they can hold. Algebra, Geometry, Calculus and many other topics can not be done using a calculator.

The real beauty and effectiveness of Vedic Mathematics cannot be fully appreciated without actually practising the system. Without practice, Vedic methods will be soon forgotten since these methods are not taught in our regular schools, so we have to consciously take time out to practise the system and it is only then that we will be benefited by these wonderful, logical, and systematic and faster methods of solving the most complex sums. One can then see that it is an extremely refined and efficient mathematical system.

Chapter 1

COMPLEMENTS

The whole number system is made up of only 9 numbers (1- 9) and a zero. All these numbers repeat themselves in a specific order after numbers like 10, 100, 1000, and so on. These numbers which have one as the first digit, followed by zeros are called **bases**.

The most interesting feature of Vedic mathematics which makes it fast and enjoyable, is the fact, that here we convert our big numbers into smaller numbers and do the calculations with them. And guess what ? We still get the same answer, as we would have from calculations made using those big numbers.

<u>Example</u>: - Instead of doing any mathematical operation from 87 we convert 87 into a small number and perform the same operation from it. Now how do we break up 87 into a smaller number? We don't randomly convert the number into any small number but we do this by subtracting 87 from its nearest base.

As 87 is close to 100 so 100 is its closest base and we subtract 87 from 100
100 – 87 = 13

This 13 is called the **Complement** of 87.
Any number when subtracted from its nearest base will give its complement.

Now if we perform any mathematical operation like multiplication, division, or any other operation by 13, we get the same answer as done by 87 in the usual way. Doesn't this sound interesting? This is what makes Vedic mathematics most **interesting, logical and fast**.

Let us try and find out complements of some numbers.
Complements of

    93 = 100 – 93 = 7
    74 = 100 – 74 = 26
    798 = 1000 – 798 = 202

We will notice that, it is easier to find the complement of smaller numbers but what do we do to find out the complement of bigger numbers? For this we have an apt sutra which comes very handy in finding the complement of any random number called

निखिलं नवतश्चरमं दशतः

"Nikhilaṃ Navataścaramaṃ Daśataḥ"

"All from nine and last from ten"

Although this sutra looks complicated, its usage is very simple. It is the most used sutra and the easiest one to understand.
Let us understand its first use with the help of a few examples

Example 1 : Complement of 743

Now instead of subtracting it from its nearest base 1000, we use the sutra.

7 4 3	We subtract each digit of our number using the sutra "All from 9 and last from 10".
From 10	Starting from the right
7 4 3↓ / 7	(1) Subtract 3 from 10 (3 is the last digit) 10 – 3 = 7 7 is the last digit of our answer.
↓From 9 7 4↓3 5 7	(2) Subtract 4 from 9 (4 is not the last digit and as the sutra says, all except last from 9) 9 – 4 = 5 5 becomes the second last digit of the answer.
↓From 9 7↓4 3 2 5 7	(3) Subtract 7 from 9 (again all digits except the last have to be subtracted from 9) 9 – 7 = 2 2 becomes the first digit of our answer.

Complement of 743 = **257**

Example 2 : Complement of 8532

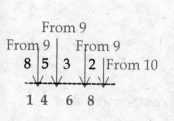

We subtract individual digits using our sutra "All from 9 last from10" from the right.
(1) 10 – 2 = 8 (2 is the last digit, so subtract it from 10)
(All other digits have to be subtracted from 9)
(2) 9 – 3 = 6
(3) 9 – 5 = 4
(4) 9 – 8 = 1
Complement of 8532 = **1468**

Example 3 : Complement of 70843

70 8 4 3
----
2 91 5 7

We subtract individual digits using our sutra "All from 9 last from10"
(1) 10 – 3 = 7 (3 is the last digit, so subtract it from 10)
Subtract all the rest from 9
(2) 9 – 4 = 5
(3) 9 – 8 = 1
(4) 9 – 0 = 9
(5) 9 – 7 = 2
Complement of 70843 = **29157**

Example 4 : Complement of 64210

6 4 2 1 0
----
3 5 7 9 0

(1) Since the last digit is 0 we just bring it down.
Why we do this is?
Because if we subtract 0 from 10 we get 10, so we will write 0 and carry over 1.
Then we subtract 1 from 9, to get 8 and add the carried over 1 and we get 9.
So it's convenient to bring the 0 down and take its next digit (1) as the last digit.
(2) 10 – 1 = 9 (1 becomes the last digit so from10)
(3) 9 – 2 = 7
(4) 9 – 4 = 5
(5) 9 – 6 = 3

Complement of   64210 = **35790**

Rules :

(1) For finding the complement of any number we subtract individual digits.

(2) Last digit has to be subtracted from 10 and all the others from 9.

(3) If the number ends in 0 then

- We just write 0 as the last digit of the complement also.

- We take the next digit from the right as the last digit and subtract it from 10.

- All the other digits have to be subtracted from 9.

Exercise 1 :

Find the complements of the following numbers :-

(1) 6213   (6) 758120

(2) 7278   (7) 925478

(3) 95102   (8) 860150

(4) 98162   (9) 862100

(5) 680054   (10) 76215000

Answers

Exercise 1
(1) 3787   (2) 2722   (3) 04898   (4) 01838
(5) 319946   (6) 241880   (7) 74521   (8) 139850
(9) 137900   (10) 23785000

Octagonal Star

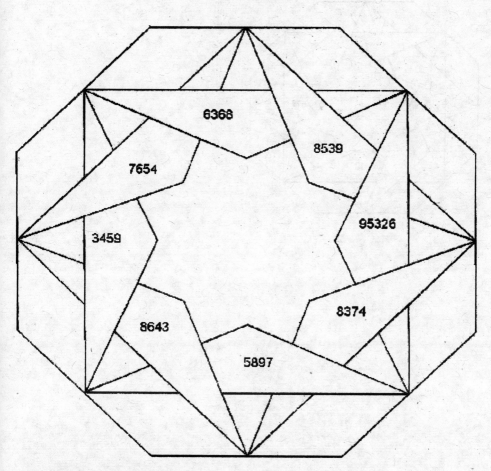

6368

8539

7654

95326

3459

8643

8374

5897

Find the complements

If the answer is:	colour the shape:
less than 3000	red
more than 3000	green

Answers to Octagonal Star

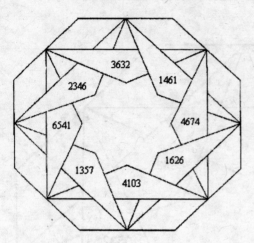

Chapter 2

SUBTRACTION

Now that we are well versed with complements, let us learn how to use them in **subtraction.**

Usually while subtracting two numbers, we write the numbers as given below. If we find the upper digit to be bigger than the lower digit we just subtract the smaller number from the bigger number.

Example :
$$
\begin{array}{r}
765 \\
- 243 \\
\hline
522 \\
\hline
\end{array}
$$

BUT when the lower digit becomes bigger than the upper digit our tedious work starts and we begin with borrowing and carry over. But this tedious task of borrowing and carry over can be avoided by using the Vedic method following the sutra:

निखिलं नवतश्चरमं दशतः

"Nikhilaṃ Navataścaramaṃ Daśataḥ"

"All from 9 and last from 10"

Let us now see how the whole calculation is carried out. As seen in the previous chapter the use of the sutra was very easy. We subtracted the last digit from ten and all other digits from nine. During subtraction we don't take the complements of the digits of our given numbers but the complement of the difference of the two digits in each column of the sum.

Let us understand it further with a few examples.

Example 1 : 723 – 246

Step 1:

```
          From 10
   7 2 3 |
 - 2 4 6 |
 --------↓
     7
 ----------
```

Here we see that in the first column,
3, the upper digit is smaller than 6, the lower digit.
We go into complements. We take the difference of
the two digits.
        6 – 3 = 3
and then the complement of the difference 3.
(Our sutra says "All from 9 and last from 10" and
since it is the first time that we are taking
complements, it is taken from 10.)

        complement of 3 from 10 is 7
        7 is the unit digit of the answer.

Step 2:

```
        From 9
   7 2 | 3
 - 2 4 | 6
 ------↓----
   7 7
 ------------
```

Here again
 2, the upper digit, is smaller than 4, the lower
digit.
We stay in complements. We take the difference of
the two digits
        4 – 2 = 2
and then the complement of 2 from 9 (Our sutra
says "All from 9 and last from 10", so this time the
complement is taken from 9.)
        complement of 2 from 9 is 7.
        7 is the tens digit of the answer.

Step 3:

```
    Sub 1 extra
   7 | 2 3
 - 2 | 4 6
 ---↓-------
   4  7 7
 -----------
```

Now
7, the upper digit, is bigger than 2, the lower digit.
So we don't need complements in this column.

Here we come out of complements.
As we are coming out of complements we subtract
the two digits and then subtract 1 extra in this
column

        7 – 2 = 5
        5 – 1 = 4 (extra for complements)
        4 is the hundreds digit of the answer.

Answer : 723 – 246 = **477**

Example 2 : 524 - 175

Step 1:

```
        From 10
  5 2 4 |
- 1 7 5 |
----------↓-
      9
------------
```

Here
    4 (upper digit) < 5 (lower digit)
so we take the complement of the difference of the two digits.
(We follow the sutra "All from 9 and last from 10" and this is the last digit so complement is taken from 10.)
      5 – 4 = 1
and then the complement of 1 from 10 = 9

      9 is the unit digit of the answer.

Step 2:

```
      From 9
  5 2| 4
- 1 7| 5
------↓-----
  4  9
-----------
```

Here again
    2 (upper digit) < 7 (lower digit)
 so we take the complement of the difference of the two digits.
(We follow the sutra "All from 9 and last from 10" and so complement is taken from 9)

    7 – 2 = 5
and the complement of 5 from 9 = 4

      4 is the tens digit of the answer.

Step 3:

```
  |Sub 1 extra
  5| 2 4
- 1| 7 5
---↓--------
  3  4 9
-----------
```

Now
    5 (upper digit) > 1 (lower digit)

so we don't need complements in this column. Here we come out of complements. As we are coming out of complements we subtract 1 extra in this column.

    5 - 1 = 4
    4 - 1 = 3 (extra for complements)

    3 is the hundreds digit of the answer.

Answer : 524 – 175 = **349**

Example 3 : 851 - 685

Step 1:

                    From 10
        8 5 1
      - 6 8 5
      ----------↓
            6

Here
    1 (upper digit) < 5 (lower digit)
so we take the complement of the difference of
the two digits.

(We follow the sutra "All from 9 and last from
10)
    5 – 1 = 4
and then the complement of 4 from 10 = 6
    6 is the unit digit of the answer.

Step 2:

                  From 9
        8 5 | 1
      - 6 8 | 5
      --------↓--
          6   6

Here again
    5 (upper digit) < 8 (lower digit)
so we take the complement of the difference of
the two digits.

    8 – 5 = 3
and the complement of 3 from 9 = 6

(We follow the sutra "All from 9 and last from
10")

6 is the tens digit of the answer.

Step 3:

      Sub 1 extra
        8 | 5 1
      - 6 | 8 5
      ------↓------
        1   6 6

Now
    8 (upper digit) > 6 (lower digit)

so we don't need complements in this column.
Here we come out of complements. As we are
coming out of complements we subtract 1 extra
in this column.

    8 – 6 = 2
    2 – 1 = 1 (extra for complements)

1 is the hundreds digit of the answer.

Answer : 851 - 635 = 166

Example 4 : 7643 – 4869

```
  7 6 4 3
- 4 8 6 9
-------------
  2 7 7 4
-------------
```

(1) 3 < 9, 9 – 3 = 6, its complement from 10 = 4
(2) 4 < 6, 6 – 4 = 2, its complement from 9 = 7
(3) 6 < 8, 8 – 6 = 2, its complement from 9 = 7
(4) 7 > 4, 7 – 4 = 3, 3 – 1 = 2 (extra for complements)

Answer : 7643 – 4869 = **2774**

Rules :

(1) When the upper digit is smaller than the upper digit then we take the complement of the difference of the two digits using the sutra "All from 9 and last from 10".

(2) When the upper digit is bigger then we come out of the complements and in this column we subtract one extra for complements.

Given below are a few more examples solving the sums in straight steps :

Example 5 : 3726 - 1987

```
  3726
- 1987
------------
  1739
------------
```

Answer : **1739**

Example 6 : 81364 - 27986

```
  81364
- 27986
--------------
  53378
--------------
```

Answer : 53378

Exercise 2 .1 :

Solve the following sums using complements :-

(1) 412 - 179

(2) 924 - 768

(3) 742 - 197

(4) 952 - 164

(5) 1000 - 872

(6) 7213 - 1796

(7) 8821 - 4992

(8) 72103 - 27984

(9) 2564 - 1697

(10) 9213 - 8954

Starting complements from the Middle of the Sum

Sometimes we find that the examples taken earlier are not the only type of subtraction problems we come across. Let us understand some other types of problems by taking a few examples.

Example 1 : 7425 – 2983

Step 1:

```
            Normal
   7 4 2 5 |
  -2 9 8 3 |
  ----------↓
         2
  ------------
```

Here

   5 (upper digit) > 3 (lower digit)

so we do normal subtraction

   5 – 3 = 2

2 is the unit digit of the answer.

Step 2:

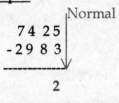

```
          From 10
   7 4 2 | 5
  -2 9 8 | 3
  ---------↓-----
       4   2
  -------------
```

Here

   2 (upper digit) < 8 (lower digit)
so we take the difference of the two digits
       8 – 2 = 6
and the complement of 6 from 10 = 4

(We follow the sutra "All from 9 and last from 10" and this is the first time we are taking complements so it is taken from 10.)

4 is the tens digit of the answer.

Step 3:

```
        From 9
   7 4 | 2 5
  -2 9 | 8 3
  --------↓-----
      4   4 2
  -----------
```

Here

   4 (upper digit) < 9 (lower digit)
so we take the difference of the two digits
       9 – 4 = 5
and the complement of 5 from 9 = 4
(We follow the sutra "All from 9 and last from 10", so from 9)

4 is the hundreds digit of the answer.

Step 4:

Sub 1 extra

```
7│4 2 5
-2│9 8 3
------
 4 4 4 2
------
```

Now
   7 (upper digit) > 2 (lower digit)
so we don't need complements in this column.
Here we come out of complements. As we are
coming out of complements we subtract 1 extra
in this column.
   7 - 2 = 5
   5 - 1 = 4 (extra for complements)
   4 is the thousands digit of the answer.

Answer : 7425 – 2983 = **4442**

Example 2 : 72354 – 19821

Step 1:

Normal

```
 72354│
-19821│
------
     3
------
```

Here
   4 (upper digit) > 1 (lower digit)

so we do normal subtraction
   4 – 1 = 3

3 is the unit digit of the answer.

Step 2 :

Normal

```
 7235│4
-1982│1
------
   3 3
------
```

Here again
   5 (upper digit) > 2 (lower digit)

so we do normal subtraction
   5 – 2 = 3

3 is the tens digit of the answer.

Step 3:

From 10

```
 723│54
-198│21
------
  5 33
------
```

Here we see that
   3 (upper digit) < 8 (lower digit)
so we take the difference of the two digits
   8 – 3 = 5
and the complement of 5 from 10 = 5
(as this is the first time we are taking
complements so it is taken from 10)

5 is the hundreds digit of the answer.

Step 4:

From 9

$$\begin{array}{r} 7\,2\,|\,3\,5\,4 \\ -1\,9\,|\,8\,2\,1 \\ \hline 2\,5\,3\,3 \\ \hline \end{array}$$

Here again

2 (upper digit) < 9 (lower digit)

so we take the difference of the two digits

9 – 2 = 7

and the complement of 7 from 9 = 2

(Rest of the complements have to be from 9)

2 is the thousands digit of the answer.

Step 5:

Sub 1 extra

$$\begin{array}{r} 7\,|\,2\,3\,5\,4 \\ -1\,|\,9\,8\,2\,1 \\ \hline 5\,|\,2\,5\,3\,3 \\ \hline \end{array}$$

Here

7 (upper digit) > 1 (lower digit)

so we don't need complements in this column so here we come out of complements. As we are coming out of complements we subtract 1 extra in this column.

7 - 1 = 6

5 - 1 (extra for complements) = 5

5 is the ten thousands digit of the answer.

Answer : 72354 – 19821 = **52533**

Example 3 : 9356 – 1794

$$\begin{array}{r} 9\,3\,5\,6 \\ -\,1\,7\,9\,4 \\ \hline 7\,5\,6\,2 \\ \hline \end{array}$$

Here

(1) 6 > 4, 6 – 4 = 2

(2) 5 < 9, 9 – 5 = 4, its complement from 10 = 6

(3) 3 < 7, 7 – 3 = 4 , its complement from 9 = 5

(4) 9 > 1, 9 – 1 = 8, 8 – 1 = 7 (extra for complement)

Answer : 9356 – 1794 = **7562**

Example 4 : 83645 - 48932

$$\begin{array}{r} 8\,3\,6\,4\,5 \\ -4\,8\,9\,3\,2 \\ \hline 3\,4\,7\,1\,3 \\ \hline \end{array}$$

(1) 5 – 2 = 3

(2) 4 – 3 = 1

(3) 9 – 6 = 3 , its complement from 10 = 7

(4) 8 – 3 = 5 , its complement from 9 = 4

(5) 8 – 4 = 4, 4 – 1 = 3 (extra for complements)

Answer : 83645 – 48932 = **34713**

Rules :

(1) Do normal subtraction when the upper digit is bigger than the lower digit.

(2) Go into complements when the lower digit is bigger than the upper digit.

(3) First time complements have to be from 10 irrespective of the column we start from.

(4) When the upper digit is bigger again come out of complements and subtract 1 extra when coming out of complements.

Given below are a few more examples solving the sums in straight steps :

Example 5 : 73594 - 58932

```
  73594
- 58932
-------------
  14662
-------------
```

Answer : **14662**

Example 6 : 538475 - 289263

```
  538475
- 289263
----------------
  249212
----------------
```

Answer : **249212**

Exercise 2 .2 :

Solve the following sums starting complements in the middle of the sum:-

(1) 7251 – 2790

(2) 3216 - 1862

(3) 7547 - 2692

(4) 6213 - 2972

(5) 1007 - 872

(6) 6294 - 1792

(7) 4201 - 2640

(8) 92138 - 27804

(9) 52798 - 26921

(10) 20454 - 19734

Leaving complements in the Middle of the Sum

Here we will take a few examples where we have bigger numbers in the upper row even after leaving complements in the sum.

Example 1 : 92173 – 60295

Step 1:

```
            From 10
   92173|
 - 60295|
 --------V
       8
 -----------
```

Here
   3 (upper digit) < 5 (lower digit)
so we go into complement

   5 – 3 = 2 , complement of 2 from 10 = 8

   8 is the unit digit of the answer.

Step 2:

```
          From 9
   9217|3
 - 6029|5
 -----V---
     7 8
 -----------
```

Here again
   7 (upper digit) < 9 (lower digit)

so 9 – 7 = 2 , complement of 2 from 9 = 7

   7 is the tens digit of the answer.

Step 3:

```
        From 9
   921|73
 - 602|95
 ----V----
    8 78
 -----------
```

Here again
   1 (upper digit) < 2 (lower digit)

so 2 – 1 = 1, complement of 1 from 9 = 8

   8 is the hundreds digit of the answer.

Step 4:

```
      Sub 1 extra
   92|173
 - 60|295
 ---V------
  1 878
 -----------
```

But here
   2 (upper digit) > 0 (lower digit)
so we come out of complement and subtract 1
extra for it

   2 – 0 = 2
   2 – 1(extra for complements) = 1

   1 is the thousands digit of the answer.

Step 5:

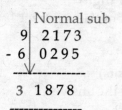

Normal sub

9 | 2173
- 6 | 0295
----▼----------

3 1878

Here again
    9 (upper digit) > 6 (lower digit)
this will be normal subtraction and since we
have already subtracted 1 extra for leaving
complements in the previous column we
DON'T have to do it again. So
    9 - 6 = 3
    3 is the ten thousands digit of the answer.

Answer : 92173 - 60295 = 31878

Example 2 : 758137 – 234792

Step 1:

          Normal
  758137 |
 -234972 |
 ----------▼

          5

Here
    7 (upper digit) > 2 (lower digit)

so 7 - 2 = 5

    5 is the unit digit of the answer.

Step 2:

          From 10
  75813 | 7
 - 23497 | 2
 -------▼----

        · 6 5

Here
    3 (upper digit) < 7 (lower digit)

so 7 – 3 = 4, complement of 4 from 10 = 6

    6 is the tens digit of the answer.

Step 3:

          From 9
  7581 | 37
 -2349 | 72
 ------▼-----

       1 65

Here
    1 (upper digit) < 9 (lower digit)

so 9 – 1 = 8, complement of 8 from 9 = 1

    1 is the hundreds digit of the answer.

Step 4:

Sub 1 extra

```
  7 5 8 | 1 3 7
- 2 3 4 | 9 7 2
----------↓--------
          3 1 6 5
```

But here

    8 (upper digit) > 4 (lower digit)

so we come out of complement and subtract 1 extra for it

    8 – 4 = 4

    4 – 1 (extra for complements) = 3

    3 is the thousands digit of the answer.

Step 5:

Normal

```
  7 5 | 8 1 3 7
- 2 3 | 4 9 7 2
---------↓-----------
      2   3 1 6 5
-----------------
```

Here

    5 (upper digit) > 3 (lower digit)

this will be normal subtraction and since we have already subtracted 1 extra for coming out of complements in the previous column we DON'T have to do it again. So

    5 – 3 = 2

    2 is the ten thousands digit of the answer

Step 6:

Normal

```
  7 | 5 8 1 3 7
- 2 | 3 4 9 7 2
--------↓----------
  5   2 3 1 6 5
-----------------
```

Here also

    7 (upper digit) > 2 (lower digit)

this will also be a normal subtraction. So

    7 – 2 = 5

    5 is the hundred thousands digit of the answer.

Answer : 75813 - 23497 = **523165**

Example 3 : 7456 – 5294

```
  7 4 5 6
- 5 2 9 4
-----------
  2 1 6 2
-----------
```

(1) 6 - 4 = 2

(2) 9 - 5 = 4, its complement from 10 = 6

(3) 4 - 2 = 2,

      2 –1 (extra for complements) = 1

(4) 7 - 5 = 2

Answer : 7456 – 5294 = **2162**

Example 4 : 53675 – 21894

$$53675$$
$$-21894$$

$$31781$$
----

(1) 5 – 4 = 1
(2) 9 – 7 = 2, its complement from 10 = 8
(3) 8 – 6 = 2, its complement from 9 = 7
(4) 3 – 1 = 2 – 1 = 1
(5) 5 – 2 = 3

Answer : 53675 – 21894 = **31781**

Rules :

(1) We start with complements when the lower digit is bigger than the upper digit irrespective of the column in the sum. Always using "All from 9 last from 10."

(2) We leave complements in the column when the upper digit is smaller than the lower digit and subtract 1 extra in that column.

(3) Once we have left the complements and have subtracted 1 extra for that. If the next upper digit is bigger we don't have to subtract an extra 1 again in that column. We carry on with it as normal subtraction.

Given below are a few more examples solving the sums in straight steps :

Example 5 : 638475 - 429763

$$638475$$
$$-429763$$

$$208712$$

Answer : **208712**

Example 6 : 4826495 - 3717871

$$4826495$$
$$-3717871$$

$$1108624$$

Answer : **1108624**

Exercise 2 .3 :

Solve the following sums leaving complements in the middle of the sum:

(1) 7521 - 2479

(2) 3520 - 1387

(3) 8654 - 2376

(4) 6524 - 1177

(5) 5273 - 1095

(6) 21748 - 10077

(7) 77326 - 52894

(8) 86249 – 23894

(9) 91021 - 60954

(10) 89625 - 63976

Colourful Cone

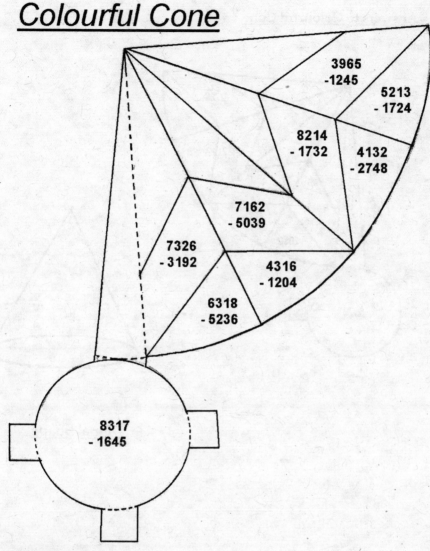

Solve the sums

If the answer is: colour the shape:
less then 3000   red
more then 3000   blue
Cut and fold it into a cone

Cut out the pattern along the outer solid line in one piece and fold along the dotted line. Tape the flaps to the underside of the cone.

Answers to Colourful Cone

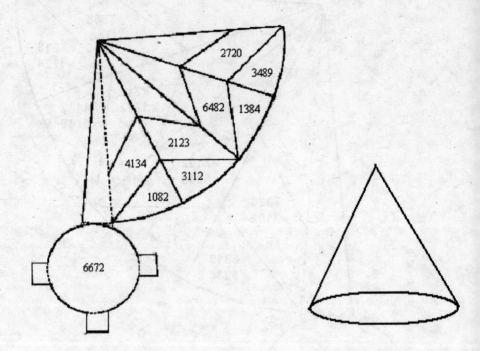

6672

2720

3489

6482 1384

2123

4134

3112

1082

Final Shape

Special Case :

There are a few special cases like subtraction of similar digits, which we often come across during subtraction, let us understand them with the help of a few examples.

Example 1 : 32868 - 17898

Step 1:

Normal

```
  32868
- 17898
-------------
       0
```

Here
8 (upper digit) = 8 (lower digit)

8 – 8 = 0
0 is the unit digit of the answer.

Step 2:

From 10

```
  3286| 8
- 1789| 8
-------------
     7 0
```

Here
6 (upper digit) < 9 (lower digit)

so 9 – 6 = 3, complement of 3 from 10 = 7

7 is the tens digit of the answer.

Step 3:

From 9

```
  328| 68
- 178| 98
-------------
    970
```

But here
8 (upper digit) = 8 (lower digit)

BUT since 8 is not bigger than 8 so we cannot come out of complements and so we take the complement of the difference.

8 – 8 = 0
complement of 0 from 9 is 9
9 is the hundreds digit of the answer.

Step 4:

From 9

```
  32| 868
- 17| 898
-------------
  4970
```

Here again
2 (upper digit) < 7 (lower digit)

so 7 – 2 = 5, complement of 5 from 9 = 4

4 is the thousands digit of the answer.

Step 5:

Sub 1 extra

```
  3 | 2868
- 1 | 7898
----↓-------
  1   4970
-------------
```

Now we see that

3 (upper digit) > 1(lower digit)

so we come out of complements

3 - 1 = 2

2 – 1(extra for complements) = 1

1 is the ten thousands digit of the answer.

Answer : 32868 – 17898 = **14970**

Example 2 : 85432 – 45496

Step 1:

From 10

```
  85432 |
- 45496 |
-----------↓
        6
------------
```

Here

2 (upper digit) < 6 (lower digit)

6 – 2 = 4, complement of 4 from 10 = **6**

6 is the unit digit of the answer.

Step 2:

From 9

```
  8543 | 2
- 4549 | 6
----------↓--
      3 6
-------------
```

Here again

3 (upper digit) < 9 (lower digit)

9 – 3 = 6 , complement of 6 from 9 is **3**

3 is the tens digit of the answer.

Step 3:

From 9

```
  854 | 32
- 454 | 96
---------↓----
    9 36
-------------
```

But here

4 (upper digit) = 4 (lower digit)

BUT since 4 is not bigger than 4 so we cannot come out of complements and so we take the complement of the difference.

4 – 4 = 0

complement of 0 from 9 is **9**

9 is the hundreds digit of the answer.

Step 4:

From 9

```
8 5 | 4 3 2
-4 5 | 4 9 6
-----↓-------
    9 9 3 6
-------------
```

But here again
  5 (upper digit) = 5 (lower digit)
  so 5 – 5 = 0
and the complement of 0 from 9 is 9

9 is the thousands digit of the answer.

Step 5:

Sub 1 extra

```
8 | 5 4 3 2
-4 | 5 4 9 6
--↓---------
3 9 9 3 6
-------------
```

Here
  8 (upper digit) > 4 (lower digit)
so we come out of complements
  8 – 4 = 4
  4 – 1 (extra for complements) = 3
  3 is the ten thousands digit of the answer.

Answer : 85432 – 45496 = **39936**

Example 3 : 53575 – 14595

```
  5 3 5 7 5
- 1 4 5 9 5
-----------
  3 8 9 8 0
-----------
```

(1) 5 – 5 = 0
(2) 9 – 7 = 2, its complement from 10 = 8
(3) 5 – 5 = 0, its complement from 9 = 9
(4) 4 – 3 = 1, its complement from 9 = 8
(5) 5 – 1 = 4 – 1 (extra for complements) = 3

Answer : 53575 – 14595 = **38980**

Example 4 : 621342 – 129371

```
  6 2 1 3 4 2
-  1 2 9 3 7 1
-------------
  4 9 1 9 7 1
-------------
```

(1) 2 – 1 = 1
(2) 7 – 4 = 3, its complement from 10 = 7
(3) 3 – 3 = 0, its complement from 9 = 9
(4) 9 – 1 = 8, its complement from 9 = 1
(5) 2 – 2 = 0, its complement from 9 = 9
(6) 6 – 1 = 5 – 1 (extra for complements) = 4

Answer : 621342 – 129371 = **491971**

Rule :

When equal numbers are subtracted outside complements they give zero, but when equal numbers are subtracted inside complements they give 9 as the complement of zero from 9 is 9.

Give below are a few more examples solving the sums in straight steps :

Example 5 : 6584537 - 3394752   **Example 6** : 54856458 - 25866948

```
    6 5 8 4 5 3 7              5 4 8 5 6 4 5 8
  - 3 3 9 4 7 5 2            - 2 5 8 6 6 9 4 8
  -----------------          --------------------
    3 1 8 9 7 8 5              2 8 9 8 9 5 1 0
  -----------------          --------------------
```

Answer : **3189785**   Answer : **28989510**

Exercise 2 .4 :

Solve the following sums leaving complements in the middle of the sum:-

(1) 72152 - 43192   (6) 28588 - 17589

(2) 92143 - 71167   (7) 72685 – 52889

(3) 621543 - 221974   (8) 56813 - 25823

(4) 752163 - 232773   (9) 91952 - 60954

(5) 52162 - 22171   (10) 65425 - 23475

Colourful Design

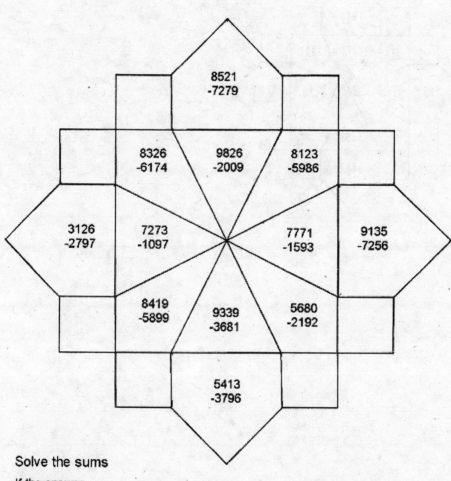

8521
-7279

8326
-6174

9826
-2009

8123
-5986

3126
-2797

7273
-1097

7771
-1593

9135
-7256

8419
-5899

9339
-3681

5680
-2192

5413
-3796

Solve the sums

If the answer
is between:

0 - 2000
2001 - 5000
5001 - 9000

Colour the shape :

Red
Blue
Yellow

Answer to the Colourful Design

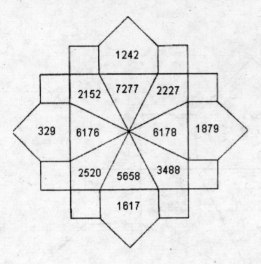

General Case

Lets us take some random examples where we need to go into complements more than once in the same sum.

Example 1 : 671245 – 380674

Step 1:

Normal
```
  671245
 -380674
---------------
        1
---------------
```

Here
5 (upper digit) > 4 (lower digit)
5 – 4 = 1
1 is the unit digit of the answer.

Step 2 :

From10
```
  67124|5
 -38067|4
---------------
      7 1
---------------
```

Here
4 (upper digit) < 7 (lower digit)

7– 4 = 3 , complement of 3 from 10 = 7

7 is the tens digit of the answer.

Step 3:

From 9
```
  6712|45
 -3806|74
---------------
    5 71
---------------
```

Here
2 (upper digit) < 6 (lower digit)

6 – 2 = 4 , complement of 4 from 9 = 5

5 is the hundreds digit of the answer.

Step 4:

```
          Sub 1 extra
   6 7 1 | 2 4 5
 - 3 8 0 | 6 7 4
 ---------↓---------
       0   5 7 1
 -----------------
```

Here
    1 (upper digit) > 0 (lower digit)
    so we come out of complements
    1 – 0 = 1
    1 – 1 (extra for complements) = 0
    0 is the thousands digit of the answer.

Step 5:

```
         | From 10
   6 7 | 1 24 5
 - 3 8 | 0 6 7 4
 ----↓---------
     9   0 5 7 1
 -----------------
```

Here again
    7 (upper digit) < 8 (lower digit)
so we again go into complements

   Here again we go into complements, and as we have already come out of complements and now we are again starting complements, so it has to be from 10 again.

    8 - 7 = 1, complement of 1 from 10 = 9
    9 is the ten thousands digit of the answer.

Step 6 :

```
          Sub 1 extra
   6 | 7 1 2 4 5
 - 3 | 8 0 6 7 4
 --↓----------
   2   9 0 5 7 1
 -----------------
```

Here
    6 (upper digit) > 3 (lower digit)
so here we come out of complements
    6 – 3 = 3
    3 – 1(extra for complements) = 2
    2 is the hundred thousands digit of the answer.

    Answer : 671245 – 380674 = **290571**

Example 2 : 569153 – 389439

Step 1:

```
   5 6 9 1 5 3
 - 3 8 9 4 3 9
 ---------------
   1 7 9 7 1 4
 ---------------
```

(1) 9 - 3 = 6, its complement from 10 = 4
(2) 5 - 3 = 2 – 1, (extra for complements) = 1
(3) 4 - 1 = 3 , its complement from 10 = 7
    (we go into complements again here)
(4) 9 - 9 = 0, its complement from 9 = 9
(5) 8 - 6 = 2, its complement from 9 = 7
(6) 5 - 3 =2 –1 (extra for complements) = 1

Answer : 569153 – 389439 = **179714**

Example 3 : 6253575 – 2821595

6253575	(1) 5 – 5 = 0
- 2821595	(2) 9 – 7 = 2, its complement from 10 = 8
---------------	(3) 5 – 5 = 0, its complement from 9 = 9
3431980	(4) 3 – 1 = 2 –1, (extra for complements) = 1
---------------	(5) 5 – 2 = 3
	(6) 8 – 2 = 6, its complement from 10 = 4
	(7) 6 – 2 = 4 – 1, (extra for complements) = 3

Answer : 6253575 – 2821595 = **3431980**

General Rules :

(1) Go into complements when the upper digit is smaller than the lower digit.

(2) For complement always use "All from 9 and last from 10."

(3) Come out of complements when the upper digit is bigger than the lower.

(4) Subtract 1 extra in the column while coming out of complements.

(5) Each time we begin complements even in the same sum first time it is from 10 and the rest from 9.

(6) When equal numbers are subtracted inside complements the answer is a 9.

Given below are a few more examples solving the sums in straight steps :

Example 4 : 5495765 - 2794389   **Example 5** : 6937841 - 3089390

```
  5495765                                    6937841
- 2794389                                  - 3089390
---------------                            -----------------
  2701376                                    3848451
---------------                            -----------------
```

Answer : **2701376**   Answer : **3848451**

When we compare both the methods we find that the Vedic method makes our calculations very short.

53623 – 37894

Usual method	Vedic method
4̶5̶ 12̶3̶ 15̶6̶ 11̶2̶ 13	53623
- 3 7 8 9 4	- 37894
-------------------	-----------
1 5 7 2 9	15729
-------------------	-----------

Exercise 2 .5 :

Solve the following subtraction sums:-

(1) 832764 - 171397

(2) 963214 - 291873

(3) 621465 - 480276

(4) 217829 – 9183

(5) 462142 - 191806

(6) 7113294 – 2009737

(7) 6539253 – 1984692

(8) 72956 – 49349

(9) 671245 - 380674

(10) 948134 - 419918

Answers

Exercise 2.1

(1) 233	(2) 156	(3) 545	(4) 788
(5) 128	(6) 5417	(7) 3829	(8) 44119
(9) 867	(10) 259		

Exercise 2.2

(1) 4461	(2) 1354	(3) 4855	(4) 3241
(5) 135	(6) 4502	(7) 1561	(8) 64334
(9) 25877	(10). 720		

Exercise 2.3

(1) 5042	(2) 2133	(3) 6278	(4) 5347
(5) 4178	(6) 11671	(7) 24432	(8) 62355
(9) 30067	(10) 25649		

Exercise 2.4

(1) 28960	(2) 20976	(3) 399569	(4) 519390
(5) 29991	(6) 10999	(7) 19796	(8) 30990
(9) 30998	(10) 41950		

Exercise 2.5

(1) 661367	(2) 671341	(3) 141189	(4) 208646
(5) 270336	(6) 5103557	(7) 4554561	(8) 23607
(9) 290571	(10) 528216		

Chapter – 3

MULTIPLICATION BY SPECIFIC NUMBERS

In Vedic methods, there are some easy ways of multiplying numbers by specific numbers, which make the whole multiplication process a very interesting one. In this chapter we will deal with some such numbers.

Multiplication by 11

We use the sutra

अन्त्ययोरेव

"Antyayoreva"

"Only the last two digits"

The application of this sutra is very easy. We just have to add only the last two digits. By using this sutra we are doing addition instead of multiplication and we get the same answer as we would have got if we had multiplied the number by 11.

Let us try and understand this with the help of a few examples.

Example 1 : 4523 x 11

Step 1

$\emptyset\ 4523\emptyset$

We place a zero on both ends of our number and convert it into a zero sandwich. We draw a line below the number where we will write the answer.

Step 2:

Ø 4 5 2 3 Ø

        3

We just keep adding the last two digits.
0 is the last digit and 3 is the second last digit.
$$0 + 3 = 3$$

Step 3:

Ø 4 5 2 3 Ø
----
      5 3

Now 3 becomes the last digit
and 2 becomes the second last digit,
so we add
$$3 + 2 = 5$$

Step 4:

Ø 4 5 2 3 Ø
----
    7 5 3

Now 2 becomes the last digit
and 5 becomes the second last digit,
so we add
$$2 + 5 = 7$$

Step 5:

Ø 4 5 2 3 Ø
----
  9 7 5 3

Now 5 becomes the last digit
and 4 becomes the second last digit,
so we add
$$5 + 4 = 9$$

Step 6:

Ø 4 5 2 3 Ø
----
4 9 7 5 3

Now 4 becomes the last digit
and 0 becomes the second last digit,
so we add
$$4 + 0 = 4$$

Answer : $4523 \times 11 = 49753$

In the Vedic method we just make a zero sandwich and keep adding the last two digits.

Example 2 : 35612×11

Step 1

Ø 3 5 6 1 2 Ø

We place a zero on both the sides of our number and convert it into a zero sandwich. We draw a line below the number where we will write the answer.

Step 2:

Ø 3 5 6 1 2 Ø

            2

We just keep adding the last two digits.
0, is the last digit and 2 is the second last
digit.
$$0 + 2 = 2$$

Step 3:

Ø 3 5 6 1 2 Ø
----
          3 2

Now 2 becomes the last digit
and 1 becomes the second last digit,
so we add
$$2 + 1 = 3$$

Step 4:

Ø 3 5 6 1 2 Ø
----
        7 3 2

Now 1 becomes the last digit
and 6 becomes the second last digit,
so we add
$$1 + 6 = 7$$

Step 5:

Ø 3 5 6 1 2 Ø

     ₁1 7 3 2
     ↵

Now 6 becomes the last digit
and 5 becomes the second last digit,
so we add
$$6 + 5 = 11$$
We place 1 as the next digit of the answer and
take the other 1 as *carry over* to the next step.

Step 6:

Ø 3 5 6 1 2 Ø
----
     9 1 7 3 2

Now 5 becomes the last digit
and 3 becomes the second last digit,
so we add
$$5 + 3 = 8$$
$$8 + 1 \text{ (carry over)} = 9$$

Step 7 :

Ø 3 5 6 1 2 Ø
----
   3 9 1 7 3 2

Now 3 becomes the last digit
and 0 becomes the second last digit,
so we add
$$3 + 0 = 3$$

Answer : 35612 × 11 = **391732**

Example 3 : 37568 × 11

Step 1

∅ 3 7 5 6 8 ∅

We convert our number into a zero sandwich and write our answer below the line.

Step 2:

⌢⌢⌢⌢⌢⌢
∅ 3 7 5 6 8 ∅
----
4 1 ₁3 ₁2 ₁4 8
⌣   ⌣   ⌣

We just keep adding the last two digits.
(1) 0 + 8 = 8
(2) 8 + 6 = ₁4
(3) 6 + 5 = 11
      11 + 1 (*carry*) = ₁2
(4) 5 + 7 = 12
      12 + 1 (*carry*) = ₁3
(5) 7 + 3 = 10
      10 + 1 (*carry*) = ₁1
(6) 3 + 0 = 3
      3 + 1 (*carry*) = 4

Answer : 37568 × 11 = **413248**

Rules :

(1) Make a zero sandwich.

(2) Add the last two digits.

(3) If we get a double-digit answer, we write down the unit digit and take the tens digit as carry over to the next step.

Given below are a few more examples solving the sums in straight steps :

Example 4 : 36142 x 11

∅ 3 6 1 4 2 ∅
----
3 9 7 5 6 2

Answer : **397562**

Example 5 : 795213 x 11

∅ 7 9 5 2 1 3 ∅
----
8 ₁7 ₁4 7 3 4 3

Answer : **8747343**

Example 6 : 618493 × 11   **Example 7 :** 179362 × 11

Ø 618493Ø   Ø 179362Ø
-------------------   -------------------
6 8 ₁0 ₁3 ₁4 ₁2 3   1 9 ₁7 ₁2 9 8 2
-------------------   -------------------

Answer : **6803423**   Answer : **1972982**

When we compare both the methods we find that the Vedic method makes our calculations very short.
4523 × 11

<u>Usual method</u>   <u>Vedic method</u>

    4523   Ø 4523 Ø
   × 11   ----------------
----------   49753
    4523
  45230

  49753

<u>Exercise 3 .1 :</u>

Multiply the following numbers by 11 :-

   (1) 52 × 11   (6) 7462 × 11

   (2) 29 × 11   (7) 2146 × 11

   (3) 314 × 11   (8) 59217 × 11

   (4) 426 × 11   (9) 217356 × 11

   (5) 213 × 11   (10) 357689 × 11

Multiplication by 12

We use the sutra

सोपान्त्यद्वयमन्त्यं

"Sopāntyadvayamantyam"

"The ultimate and twice the penultimate"

While following this sutra here too we are doing addition instead of multiplication and we still get the same answer as we would get if we had multiplied the number by 12. Like the previous multiplication here also we make a zero sandwich and then just add the ultimate (last) digit and **twice** the penultimate (second last) digit.

Let us understand this with the help of a few examples:

Example 1 : 124 × 12

Step 1

$\varnothing$ 1 2 4 $\varnothing$

We place a zero on both ends of our number and convert it into a zero sandwich. We draw a line below the number where we will write the answer.

In this case we keep adding the last digit and twice the second last digit.

Step 2:

$\varnothing$ 1 2 4 $\varnothing$

      8

Here the last digit is 0 and the second last digit 4.

We add the last digit (0) plus twice the second last digit (4)
$$0 + 2 \times 4 = 8$$

Step 3:

We add the last digit (4) plus twice the second last digit (2)

$$4 + 2 \times 2 = 8$$

Ø 1 2 4 Ø
----
8 8

Step 4:

Ø 1 2 4 Ø
----
4 8 8

We add the last digit (2) plus twice the second last digit (1)
$$2 + 2 \times 1 = 4$$

Step 5:

Ø 1 2 4 Ø
----
1 4 8 8

We add the last digit (1) plus twice the second last digit (0)
$$1 + 2 \times 0 = 1$$

Answer : 124 × 12 = 1488

Example 2 : 26154 × 12

Step 1

Ø 2 6 1 5 4 Ø

We place a zero on both the sides of our number and convert it into a zero sandwich. We draw a line below the number where we will write the answer.

Here we keep adding the last digit and twice the second last digit.

Step 2:

Ø 2 6 1 5 4 Ø

8

We add the last digit 0 plus twice the second last digit 4

$$0 + 2 \times 4 = 8$$

Step 3:

Ø 2 6 1 5 4 Ø

₁4 8

We add the last digit (4) plus twice the second last digit (5)

$$4 + 2 \times 5 = {}_1 4$$

We place 4 as the next digit of the answer and take the other 1 as *carry over* to the next step.

Step 4:

Ø 26154Ø

     848

We add the last digit (5) plus twice the second last digit (1)

$$5 + 2 \times 1 = 7$$
$$7 + 1 \ (carry) = 8$$

Step 5:

Ø 26154Ø

  ₁3848

  ↵

We add the last digit (1) plus twice the second last digit (6)

$$1 + 2 \times 6 = {}_13$$

We place 3 as the next digit of the answer and take the other 1 as *carry over* to the next step.

Step 6 :

Ø 26154Ø

 ₁13848

↵

We add the last digit (6) plus twice the second last digit (2)

$$6 + 2 \times 2 = 10$$
$$10 + 1 \ (carry) = {}_11$$

We place 1 as the next digit of the answer and take the other 1 as *carry over* to the next step.

Step 7 :

Ø 26154Ø

313848

We add the last digit (2) plus twice the second last digit (0)

$$2 + 2 \times 0 = 2$$
$$2 + 1 \ (carry) = 3$$

Answer : 26154 × 12 = **313848**

Example 3 : 82643 × 12

Step 1

Ø 82643Ø

We convert our number into a zero sandwich and write our answer below the line.

Step 2:

$\sim\sim\sim\sim\sim$
Ø 8 2 6 4 3 Ø
----
9 ₁9 ₁1 ₁7 ₁1 6

(1) $0 + 2 \times 3 = 6$
(2) $3 + 2 \times 4 = {}_11$
(3) $4 + 2 \times 6 = 16$
  $16 + 1 \ (carry) = {}_17$
(4) $6 + 2 \times 2 = 10$
  $10 + 1 \ (carry) = {}_11$
(5) $2 + 2 \times 8 = 18$
  $18 + 1 \ (carry) = {}_19$
(6) $8 + 2 \times 0 = 8$
  $8 + 1 \ (carry) = 9$

Answer : $82643 \times 12 = 991716$

Rules :

(1) Make a zero sandwich.

(2) Add the last and twice the second last digit.

(3) If we get a double-digit answer we put down the unit digit and carry over the tens digit to the next step.

Given below are a few more examples solving the sums in straight steps :

Example 4 : 4137 x 12

Ø 4 1 3 7 Ø
----
4 9 6 ₁4 ₁4

Answer : **49644**

Example 5 : 713624 x 12

Ø 7 1 3 6 2 4 Ø
----
8 ₁5 6 ₁3 ₁4 8 8

Answer : **8563488**

Example 6 : 3176214 x 12

Ø 3 1 7 6 2 1 4 Ø
----
3 8 ₁1 ₂1 ₁4 5 6 8

Answer : **38114568**

Example 7 : 416295 x 12

Ø 4 1 6 2 9 5 Ø
----
4 9 9 ₁5 ₁5 ₂4 ₁0

Answer : **4995540**

When we compare both the methods we find that the Vedic method makes our calculations very short.

124 x 12

Usual method	Vedic method
124	Ø 124 Ø
× 12	-------------
--------	1488
248	-------------
1240	

1488	

Exercise 3 .2 :

Multiply the following numbers by 12 :-

(1) 56 x 12   (6) 2134 x 12

(2) 42 x 12   (7) 5162 x 12

(3) 78 x 12   (8) 71213 x 12

(4) 124 x 12   (9) 21968 x 12

(5) 639 x 12   (10) 2701215 x 12

Fisherman's Boat

Solve the problems and connect
their dot to the dot beside their
answer.

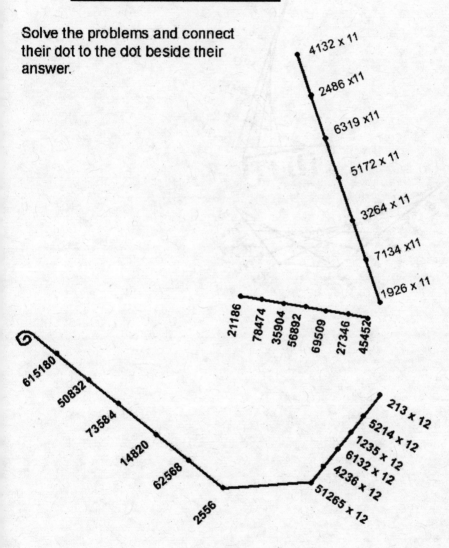

4132 x 11

2486 x11

6319 x11

5172 x 11

3264 x 11

7134 x11

1926 x 11

21186
78474
35904
56892
69509
27346
45452

615180
50832
73584
14820
62568
2556

213 x 12
5214 x 12
1235 x 12
6132 x 12
4236 x 12
51265 x 12

Answers to the Fisherman's Boat

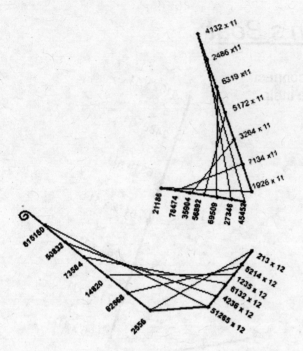

Multiplication by 9

In the whole number series 9 is the most interesting number therefore the multiplication of 9 is also very interesting. If we have a number to be multiplied by a number consisting of only 9's we use the sub sutra

एकन्यूनेन पूर्वेन

"Ekanyūnena Pūrvena"

"By one less than the one before"

along with our sutra

निखिलं नवतश्चरमं दशतः

"Nikhilaṃ Navataścaramaṃ Daśataḥ"

"All from nine and last from ten"

We have categorized the multiplication by 9 in three types. Let us understand all the three types one by one.

First Type :

Where the number of digits in the multiplicand (number to be multiplied) and the multiplier are the same and all the digits of the multiplier are 9's.

<u>**Example 1**</u> : 342 x 999

Step 1:   We divide our answer in two parts, the Left
                           Hand Side (L.H.S.) and the Right Hand Side
       341 /   (R.H.S.)

                           <u>L.H.S.</u>: We write one less than the
                           multiplicand.
                                    342 – 1 = 341

Step 2:   <u>R.H.S.</u>: We write the complement of the
                           multiplicand using "All from 9 last from
    341 / 658   10"sutra.

                           complement of 342 = 658

                           OR subtract each digit of the L.H.S. of the
                           answer (341) individually from 9.
                           (9 - 3 = 6), (9 – 4 = 5) , (9 – 1 = 8) = 658

                           Answer : 342 × 999 = **341658**

<u>**Example 2**</u> : 536874 x 999999

Step 1:   We divide our answer in two parts.

   536873 /   <u>L.H.S.</u>: We write one less than the
                           multiplicand
                                    536874 – 1 = 536873

Step 2:   <u>R.H.S.</u>: We write the complement of the
                           multiplicand using "All from 9 last from
536873 / 463126   10"sutra.

                           complement of 536874 = 463126
                           OR subtracting each digit of the LHS of the
                           answer from 9 we also get 463126

                           Answer : 536874 × 999999 = **536873463126**

Example 3 : 2038405 x 9999999

Step 1:   We divide our answer in two parts.

2038404 /   **L.H.S.:** We write one less than the
                                            multiplicand
                                            2038405 – 1 = 2038404

Step 2:   **R.H.S.** : complement of 2038405 = 7961595

2038404 / 7961595   **OR** subtracting each digit of the L.H.S. of the
                                            answer from 9 we also get 7961595

                                            Answer : 2038405 x 9999999
                                            = 20384047961595

Rules :

(1) We get the answer in 2 steps.

(2) Left hand side of the answer is one less than the multiplicand
(number to be multiplied).

(3) Right hand side is the complement of the multiplicand number or
we can subtract each digit of the left hand side of the answer
individually from 9.

(4) The most important to remember is that the digits in the
multiplicand should be equal to the digits in the multiplier.

Given below are a few more examples solving the sums in straight steps :

Example 4 :   **Example 5** :

5642 x 9999 = 5641 / 4358   71496 x 99999 = 71495 / 28504

Answer : 56414358   Answer : 7149528504

Example 6 :

1745213 x 9999999

= 1745212 / 8254787

Answer : 17452128254787

Example 7 :

3417629 x 9999999

= 3417628 / 6582371

Answer : 34176286582371

When we compare both the methods we find that the Vedic method makes our calculations very short.

475 x 999

Usual method	Vedic method
475	475 x 999
x 999	= 474 / 525
----------	= 474525
4275	
42750	
427500	

474525	

Exercise 3 .3 :

Solve the following sums mentally :-

(1) 46 × 99

(2) 74 × 99

(3) 65 × 99

(4) 694 × 999

(5) 374 × 999

(6) 618 × 999

(7) 6842 × 9999

(8) 98246 × 99999

(9) 654972 × 999999

(10) 68245731 × 99999999

Second Type:

In this case the numbers of digit in the multiplier are more than the digit in the multiplicand.

This is also equally easy as the previous one. We just have to add zero's before the multiplicand and make the number of digits equal to the digits of the multiplier.

Example 1 : 23 x 999

Step 1:

023 x 999

Since the multiplicand had 2 digits and multiplier has 3 digits we make our number 023, so that the digits of the multiplicand are equal to the digits of the multiplier.
Now we divide our answer in two parts.

Step 2:

22 /

L.H.S.: We write one less than the multiplicand

$$23 - 1 = 22$$

Step 3:

22 / 977

R.H.S. : The complement of 023 = 977

OR subtracting each digit of the L.H.S. of the answer from 9 we also get 977

Answer : 23 x 999 = **22977**

Example 2 : 548 x 99999

Step 1:

00548 x 99999

Since the multiplicand has 3 digits and multiplier has 5 digits.

we convert our number to 00548
so that the digits on both the multiplier and multiplicand are equal.
Now we divide our answer in two parts.

Step 2:

547 /

L.H.S.: We write one less than the multiplicand
$$548 - 1 = 547$$

<u>Step 3:</u>   <u>R.H.S.</u> : The complement of 00548 = 99452

547 / 99452   OR subtracting each digit of the L.H.S. of the
                                 answer from 9 we also get 99452

                                 Answer : 548 x 99999 = **54799452**

Example 3 : 7825 x 9999999

<u>Step 1:</u>   We convert our number to 0007825

0007825 x 9999999   Now we divide our answer in two parts.

<u>Step 2:</u>   <u>L.H.S.</u> :
                                      7825 – 1 = 7824
7824 /

<u>Step 3:</u>   <u>R.H.S.</u> :
                                 The complement of 0007825 = 9992175
7824 / 9992175
                                 OR subtracting each digit of the L.H.S. of the
                                 answer from 9 we also get 9992175

                                 Answer : 7825 x 9999999 = **78249992175**

Given below are a few more examples solving the sums in straight steps :

Example 4 :   **Example 5 :**

742 x 9999 = 741 / 9258   5761 x 999999 = 5760 / 994239

Answer : **7419258**   Answer : **5760994239**

Example 6 :

32465 x 9999999
= 32464 / 9967535

Answer : **324649967535**

Example 7:

46314 x 99999999
= 46313 / 99953686

Answer : **4631399953686**

We also observe from the above examples that, the number of extra nine's in the multiplier (the number of digits in the multiplier more than the multiplicand), is the number of nine's present in the middle of the answer. The answer can be written in three parts as:

(One less than the multiplicand)/ (No. of nines extra)/ (complement of the multiplicand)

Example : 23 x 999 = 22 / 9 / 77 = 22977

6547 x 999999 = 6546 / 99 /3453 = 6546993453

Rules :

(1) Make the digits of the multiplicand equal to the digits of the multiplier by adding zeros before the number.

(2) The Left hand side of the answer is one less than the number.

(3) The Right hand side of the answer is the complement of the number (the one formed by adding the zeros).

(4) Or place the extra number of nines in the middle of both the left and the right part of the answer, and then write the complement of the original multiplicand.

Exercise 3 .4 :

Solve the following sums mentally :-

(1) 93 × 999   (6) 2564 × 99999

(2) 76 × 9999   (7) 3654 × 99999

(3) 543 × 9999   (8) 289 × 999999

(4) 324 × 99999   (9) 26845 × 99999999

(5) 872 × 99999   (10) 53167 × 99999999

Third Type :

In this case the number of digits in the multiplier are LESS than the number of digits in the multiplicand but here also all the digits in the multiplier should be 9's.

Example 1 : 436 x 99

<u>Step 1:</u> 4: 36	We divide our multiplicand into two parts such that the right side has the same number of digits as the number of 9's in the multiplier. We have two 9's in the multiplier so two digits in the right part. So we have 4 in the left part and 36 in the right part We separate the two parts with a colon.

<u>Step 2:</u>
 	
  4 : 36
   - 5
= 431 /
	L.H.S. :
	We add 1 to the left part
	     4 + 1= 5
	and subtract it from the whole multiplicand.
	    436 – 5 = 431

Step 3 :

4 : 36

= 431 / 64

R.H.S. :
We write the complement of the right part of the multiplicand next to it.
The complement of 36 = 64

Answer : 436 x 99 = 43164

Example 2 : 20462 x 999

Step 1:

20 : 462

We divide our multiplicand in two groups such that the right side has the same number of digits as the number of 9's in the multiplier.
We have three 9's in the multiplier so three digits in the right part.

So we have 20 in the left part and
462 in the right part

Step 2:

20 : 462
- 21
= 20441 /

L.H.S. :
We add 1 to the left part
20 + 1= 21
and subtract it from the whole multiplicand.

20462 – 21 = 20441

Step 3:

20 : 462

= 20441 / 538

R.H.S. :
The complement of 462 (right part of the multiplicand)
= 538

Answer : 20462 x 999 = 20441538

Example 3 : 46972 x 99

Step 1:

469 : 72

We divide our multiplicand in two groups such that we have two 9's in the multiplier so two digits in the right part.

We have 469 in the left part and
72 in the right part

<u>Step 2:</u>   <u>L.H.S. :</u>

469 + 1 = 470

469 : 72   46972 – 470 = 46502
   – 4 70

           <u>R.H.S. :</u>

= 46502 / 28   The complement of 72 = 28

Answer : 46972 x 99 = 4650228

<u>Rules :</u>

(1) We divide our multiplicand in two parts such that the right side has the same number of digits as the number of 9's in the multiplier.

(2) We increase the left part by one and subtract it from the multiplicand to form the Left hand side of the answer.

(3) The Right hand side of the answer is found by taking the complement of the right part of the multiplicand.

Given below are a few more examples solving the sums in straight steps :

Example 4 : 462 x 99   **Example 5** : 25473 x 99

    4 : 62   254 : 73
     – 5   – 255
   -----------   ---- -----------
   457 / 38   25218 / 27

Answer : **45738**   Answer : **2521827**

Example 6 : 21735 x 999   **Example 7** : 521367 x 999

   21 : 735   521 : 367
    – 22   – 522
  -----------------   ------------------
  21713 / 265   520845 / 633

Answer : **21713265**   Answer : **520845633**

When we compare both the methods we find that the Vedic method makes our calculations very short.

4937 x 9999

<table>
<tr><td>Usual method</td><td>Vedic method</td></tr>
<tr><td>4937</td><td>4937 x 9999</td></tr>
<tr><td>x 9999</td><td>= 49365063</td></tr>
</table>

Usual method
$$4937$$
$$\times 9999$$

$$44433$$
$$444330$$
$$4443300$$
$$44433000$$

$$49365063$$

Exercise 3 .5 :

Solve the following sums:-

(1) 48 × 9

(2) 67 × 9

(3) 122 × 9

(4) 359 × 99

(5) 4599 × 99

(6) 568 × 9

(7) 1253 × 999

(8) 59731 × 999

(9) 25999 × 999

(10) 25869 × 99

Answers

Exercise 3 .1

(1) 572	(2)319	(3) 3454	(4) 4686
(5) 57343	(6) 80282	(7) 23606	(8) 651387
(9) 2390916	(10) 3934579		

Exercise 3 .2

(1) 672	(2) 504	(3) 936	(4) 1488
(5) 7668	(6) 25608	(7) 61944	(8) 854556
(9) 263616	(10) 32414580		

Exercise 3 .3

(1) 4554	(2) 7326	(3) 6435	(4) 693306
(5) 373626	(6) 617382	(7) 68413158	(8)9824501754
(9) 654971345028	(10) 6824573031754269		

Exercise 3 .4

(1) 92907	(2) 749924	(3) 5429457	(4) 32399676
(5) 87199128	(6) 256397436	(7) 365396346	(8) 288999711
(9) 2684499973155	(10) 5316699946833		

Exercise 3 .5

(1) 432	(2) 603	(3) 1098	(4) 35541
(5) 455301	(6) 5112	(7) 1251747	(8) 59671269
(9) 25973001	(10) 2561031		

Chapter 4

BASE MULTIPLICATION

We have different methods of multiplication for different types of numbers. In the previous chapter we studied multiplication by specific numbers here we will discuss multiplication of numbers close to the bases (which have a small complement). We know that bases are the numbers which have 1 followed by zeros, like 10, 100, 1000, etc.

For this type of multiplication we use the sutra

निखिलं नवतश्चरमं दशतः

"Nikhilam Navataścaramam Daśataḥ"

"All from nine and last from ten"

If the numbers to be multiplied are close to any one base, either more or less, then our multiplication becomes very easy and very fast. Both the numbers can either be more than the base or both can be less or one can be more and one less than the base but both the numbers should have one common base.

First we will discuss multiplication of numbers where *both the numbers are less than the base.* Or both the numbers have a deficiency from the base.

Let us try and understand it with a few examples

Example 1 : 88 × 98

Step 1:

Here both the numbers are close to the base 100, so we write our given numbers as shown.

```
  100
----------
   88
   98
----------
```

Step 2 :

100
----
88 - 12
.98

We know that 88 is 12 less than the base or its complement is 12.

so we write -12 (minus 12) next to 88 as our given numbers are less than the base.

Step 3:

100
----
88   - 12
98   - 02

We know that 98 is 2 less than the base or its complement is 02.

so we write – 02 (minus 02) next to 98.

We will get our answer in two parts, Right hand side (R.H.S.) and Left hand side (L.H.S.)

Step 4 :

100
----
88   - 12 | multiply
98   - 02 |
------------↓
/ 24

R.H.S. :
We multiply the two complements of both our numbers.
$$(- 12) \times (- 02) = 24$$

Since we have base as 100 in this sum and 100 has 2 zeros so the R.H.S. must have only 2 digits.
And here also we have 2 digits.

Step 5 :

100
----
88 ⤬ - 12 cross
98   - 02 subtract
----
86 / 24

L.H.S. :
We can get our answer in 4 different ways.
(1) Cross subtraction : 88 - 02 = 86
(2) Cross subtraction : 98 - 12 = 86
(3) Add the two given numbers and subtract the base from it :
$$(88 + 98) – 100 = 186 – 100 = 86$$
(4) Subtract both the complements from the base :
$$100 -12 - 2 = 86$$

We get the same answer irrespective of the method we choose. So we can choose any one method.

Answer : 88 × 98 = 8624

Example 2 : 97 × 98

Step 1:

```
     100
  -------------
  97
  98
  -------------
```

Here both the numbers are close to the base 100, so we write our given numbers as shown.

Step 2 :

```
     100
  -------------
  97      - 03
  98
  -------------
```

We know that 97 is 3 less than the base or its complement is 3.

so we write - 03 (minus 03) next to 97.

Step 3:

```
     100
  -------------
  97      - 03
  98      - 02
  -------------
```

We know that 98 is 2 less than the base or its complement is 2.

so we write – 02 (minus 02) next to 98.

We will get our answer in two parts.

Step 4 :

```
     100
  -------------
  97      - 03  |multiply
  98      - 02  ↓
  -------------
        /  06
  -------------
```

R.H.S. :
We multiply the complements of both our given numbers.
$$(- 03) × (- 02) = 6$$

Since we have base as 100 in this sum and 100 has 2 zero so the R.H.S must have only 2 digits.
Here we have 1 digit so we write
6 as 06 to make it 2 digits.

Step 5 :

```
     100
  -------------
  97  ⤬ - 03   cross
  98      - 02  subtract
  -------------
  95  /  06
```

L.H.S. :
We can get our answer in 4 different ways.
(1) Cross subtraction : 97 - 02 = 95
(2) Cross subtraction : 98 - 03 = 95
(3) Add the two given numbers and subtract the base from it :
$$(97 + 98) – 100 = 195 – 100 = 95$$

(4) Subtract both the complements from the base :

$$100 - 03 - 02 = 95$$

We get the same answer irrespective of the method we choose.

Answer : $97 \times 98 = 9506$

Example 3 : 91×85

Step 1:

```
  100
-----------
  91
  85
-----------
```

Here both the given numbers are close to the base 100, so we write our numbers as shown.

Step 2 :

```
  100
-----------
  91   - 09
  85
-----------
```

We know that 91 is 9 less than the base or its complement is 9.

so we write - 09 next to 91

Step 3:

```
  100
-----------
  91   - 09
  85   - 15
-----------
```

We know that 85 is 15 less than the base or its complement is 15.

so we write – 15 next to 85.

We will get our answer in two parts.

Step 4 :

```
  100
-----------
  91   - 09 | multiply
  85   - 15 |
-----------  ↓
     / ₁3 5
------- ⋃ -----
```

R.H.S. :

We multiply the two complements of both our given numbers.

$(- 09) \times (- 15) = {}_1 35$

Since we have base as 100 in this sum and 100 has 2 zeros so the R.H.S. must have only 2 digits and here we have 3 digits.

So we write **35** in right side and *carry 1* to left side.

Step 5 :

```
        100
    ---------------
    91      - 09   cross
    85      - 15 subtract
    ---------------
    77  /  35
    ---------------
```

L.H.S. :

We can get our answer in 4 different ways.

(1) Cross subtraction : 91 – 15 = 76

(2) Cross subtraction : 85 – 09 = 76

(3) Add the two given numbers and subtract the base from it :

$$(91 + 85) – 100 = 176 – 100 = 76$$

(4) Subtract both the complements from the base :

$$100 - 09 - 15 = 76$$

We get the same answer irrespective of the method we choose. We now add the carry over from the right side to it to get the answer.

$$76 + 1 \ (carry) = 77$$

Answer : 91 × 8 5 = 7735

Example 4 : 997 × 993

Step 1:

```
      1000
    ------------
    997
    993
```

Here both the numbers are close to the base 1000, so we write our numbers as shown.

Step 2:

```
        1000
    ----------------
    997  - 003
    993
    ----------------
```

We know that 997 is 3 less than the base or its complement of 997 from 1000 is 003.

so we write - 003 next to 997.

Step 3:

```
        1000
    ----------------
    997  - 003
    993  - 007
    ----------------
```

We know that 993 is 07 less than the base or its complement is 007.

so we write – 007 next to 993.

We will get our answer in two parts.

Step 4 :
```
        1000
------------------
   997    - 003 | multiply
   993    - 007 |
----------------↓
       /  021
------------------
```

R.H.S. :
We multiply the complements of both our numbers.

$$(-003) \times (-007) = 21$$

Since we have base as 1000 in this sum and 1000 has 3 zeros so the R.H.S must have only 3 digits and here we have 2 digits.
So we write 21 as 021

Step 5 :
```
      1000
------------------
   997  ⟍↗ - 003   cross
   993  ↗⟍ - 007   subtract
------------------
   990  /  021
------------------
```

L.H.S. :
We can get our answer in 4 different ways.
(1) Cross subtraction : 997 - 007 = 990
(2) Cross subtraction : 993 - 0 3 = 990
(3) Add the two given numbers and subtract the base from it :
   (997 + 993) – 1000 = 1990– 1000 = 990
(4) Subtract both the complements from the base :
   1000 -003 - 007= 990

We get the same answer irrespective of the method we choose.

Answer : 997 × 993 = 990021

Example 5 : 884 x 996

Step 1:
```
     1000
------------------
   884  - 116
   996  - 004
------------------
```

Here both the numbers are close to the base 1000, so we write our numbers as shown.

We place their complements next to them.

We will get our answer in two parts.

Step 2:

```
      1000
--------------------
 884  ⟍ ↗  - 116
 996  ⟋ ↘  - 004
--------------------
 880   /   464
--------------------
```

R.H.S. :

    116 x 004 = 464

L.H.S. :

We can choose any one method, so the easier ones are :

    884 - 004 = 880
or 996 - 116 = 880

Answer : 884 x 996 = 880464

Example 6 : 76 x 89

Step 1:

```
      100
---------------
 76   - 24
 89   - 11
---------------
```

Here both the numbers are close to the base 100, so we write our numbers as shown.

We place their complements next to them.

We will get our answer in two parts.

Step 2:

```
      100
---------------
 76  ⟍ ↗ - 24
 89  ⟋ ↘ - 11
---------------
 67  /  ₂64
------- ↻ -------
```

R.H.S. :

    24 x 11 = ₂64

(we can use zero sandwich method)

We write 64 in the R.H.S. and *carry over* 2 to L.H.S.

L.H.S. :

We can choose any one method so

    76 – 11 = 65

or 89 - 24 = 65

    65 + 2 (*carry*) = 67

Answer : 76 x 89 = 6764

Rules :

(1) We find out the complement of the two given numbers and write them next to the numbers.

(2) Right hand side of the answer is the multiplication of the complements (deficiency, in case when the numbers are less than the base) of the numbers keeping into account the number of zeros in the base.

(3) For the left hand side, we subtract the deficiency of the first number from the second number or vice versa.

(4) Count the number of zeros in the base, the right side of the answer should have the same numbers of digits.

(5) If we have more digits we carry them over to the left side, but

(6) If we have less number of digits then we add zeros before the number to make the digits on right side equal to the number of zeros in the base.

Given below are a few more examples solving the sums in straight steps :

Example 7 : 98 x 92 –

```
       100
    -----------
    98 - 02
    92 - 08
    -----------
    90 / 16
```

Answer : **9016**

Example 8 : 89 x 78

```
       100
    -----------
    89 - 11
    78 - 22
    -----------
    69 / ₂42
```

Answer : **6942**

Example 9 : 986 x 992

```
   1000
---------------
986 – 014
992 -  008
---------------
 978  / 112
```

Answer : **978112**

Example 10 : 9927 x 9998

```
   10000
--------------
9927 – 0073
9998 – 0002
----------------
9925 / 0146
```

Answer : **99250146**

If we observe carefully we can do these types of sums mentally in the following way :-

L.H.S. : (First number MINUS the complement of the second number)
R.H.S. : (product of both the complements)

Example : 98 × 96 = (98 - 04) / (02 × 04)
               = 94 /08
               = 9408

        78 x 89 = (78 - 11)/ (22 x 11)
               = 67 / $_2$42
               = 6942

    994 × 992 = (994 - 008) / (006 × 008)
               = 986 / 048
               = 986048

When we compare both the methods we find that the Vedic method makes our calculations very short.

96 x 82

<u>Usual method</u>　　　　　　　　　　　　　　　　<u>Vedic method</u>

```
        96                        96  - 04
      x 82                        82  - 18
    -----------                 -------------
       192                        78  /  72
      7680                      -------------
    -----------
      7872
    -----------
```

<u>Exercise 4 .1 :</u>

Solve the following sums :-

(1) 99 × 98　　　　　　　　(6) 992 × 993

(2) 98 × 96　　　　　　　　(7) 998 × 882

(3) 96 × 94　　　　　　　　(8) 9995 × 9980

(4) 92 × 95　　　　　　　　(9) 9988 × 9988

(5) 999 × 996　　　　　　　(10) 9889 × 9992

Multiplication of Numbers more than the Base

Example 1 : 17 × 13

Step 1:

```
      10
   ----------
   17
   13
   ----------
```

Here both the numbers are close to the base 10, so we write our given numbers as shown.

Step 2 :

```
      10
   ----------
   17  + 7
   13
   ----------
```

We know that 17 is 7 more than the base so we write +7 (positive 7) next to 17.

Step 3:

```
      10
   ----------
   17   + 7
   13   + 3
   ----------
```

We know that 13 is 3 more than the base so

we write +3 (positive 3) next to 13.

Step 4 :

```
      10
   ------------
   17    + 7  | multiply
   13    + 3  ↓
   ------------
       / ₂1
   ---⤸------
```

We will get our answer in two parts.
R.H.S. : -
We multiply the two complements (excesses) of both our numbers.
$$(+7) × (+3) = 21$$
Since 10 is the base so we must have only 1 digit in R.H.S. We write 1 in R.H.S. and take 2 as *carry over* to left side.

Step 5 :

```
            10
    ------------------
    17  ⟍↗  + 7   cross
    13  ⟋⟍  + 3   addition
    ------------------
     22  /  1
    ------------------
```

L.H.S. :

We can get our answer in 4 different ways.

(1) Cross addition : 17 + 3 = 20

(2) Cross addition : 13 + 7 = 20

(3) Add the two numbers and subtract the base from it :
$$(17 + 13) - 10 = 30 - 10 = 20$$

(4) Subtract both the complements from the base :
$$10 + 7 + 3 = 20$$

Now   20 + 2 *(carry)* = 22

Answer : 17 × 13 = **221**

Example 2 : 104 × 112

Step 1:

```
        100
    --------------
    104   + 04
    112   + 12
    --------------
```

Here both the numbers are close to the base 100, so we write our numbers as shown.

We place their complements next to them.

We will get our answer in two parts.

Step 2 :

```
        100
    --------------
    104    + 04  │ multiply
    112    + 12  │
    --------------↓
         /  48
    --------------
```

R.H.S. :
$$(+ 4) × (+ 12) = 48$$

Since 100 is the base so we can have 2 digits.

Step 3 :

```
        100
  -------------------
  104      +04   cross
  112      +12   addition
  -------------------
   116  /   48
  -------------------
```

L.H.S. :

We can get our answer in 4 different ways.
(1) Cross addition : 104 + 12 = 116
(2) Cross addition : 112 + 4 = 116
(3) Add the two numbers and subtract the base from it :
$$(104 + 112) - 100 = 116$$
(4) Subtract both the complements from the base :
$$100 + 12 + 4 = 116$$

Answer : 104 × 112 = **11648**

Example 3 : 106 × 122

Step 1:

```
       100
  ---------------
   106  + 06
   122  + 22
  ---------------
```

Here both the numbers are close to the base 100, so we write our numbers as shown.

We place their complements next to them.

We will get our answer in two parts.

Step 2 :

```
       100
  ---------------
   106   + 06|  multiply
   122   + 22|
  ------------↓
         /  ₁32
  ------  ↻ -------
```

R.H.S. :
$$(+ 6) × (+ 22) = {}_1 32$$

Since 100 is the base so we can have 2 digits so we place 32 and *carry over* 1 to left side.

Step 3 :

```
            100
-----------------------
106    ⟋ + 06  cross
122    ⟍ + 22  addition
-----------------------
 129  /  32
-----------------------
```

L.H.S. :

We can get our answer in 4 different ways.

(1) Cross addition : $106 + 22 = 128$
(2) Cross addition : $122 + 06 = 128$
(3) Add the two numbers and subtract the base from it :
    $$(106 + 122) - 100 = 128$$
(4) Subtract both the complements from the base :
    $$100 + 22 + 06 = 128$$
    Now $128 + 1(carry) = 129$

Answer : $106 \times 122 = 12932$

Example 4 : 1067×1005

Step 1:

```
         1000
-----------------
1067   +067
1005   +005
-----------------
```

Here both the numbers are close to the base 1000, so we write our numbers as shown.

We place their complements next to them. We will get our answer in two parts.

Step 2 :

```
          1000
------------------------
1067  ⟋ + 067  |
1005  ⟍ + 005  ↓
------------------------
1072 / 335
------------------------
```

R.H.S. :
    $(+ 067) \times (+ 005) = 335$
L.H.S. :

    $1067 + 005 = 1072$
or $1005 + 067 = 1072$

Answer : $1067 \times 1005 = \mathbf{1072335}$

Rules :

(1) We find out the complement of the two given numbers and write them next to the numbers.

(2) Right hand side of the answer is the multiplication of the complement (excess in case on numbers are more than the base) of the numbers.

(3) Left hand side, we add the excess of the first number to the second number or the excess of the second number to the first number, (we will get the same answer both the ways)

(4) Count the number of zeros in the base, The Right hand side of the answer should have the same number of digits.

- If we have more number of digits we carry them over to the left side,

- If we have less number of digits then we add zeros before the number to make the digits on right side equal to the number of zeros in the base.

Given below are a few more examples solving the sums in straight steps :

Example 5 : 106 x 111

$$100$$

$$106 + 06$$
$$111 + 11$$

$$117 / 66$$

Answer : **11766**

Example 6 : 1042 x 1002

$$1000$$

$$1042 + 042$$
$$1002 + 002$$

$$1044 / 084$$

Answer : **1044084**

Example 7 : 121 x 108

$$100$$

$$121 + 21$$
$$108 + 08$$

$$130 /_1 68$$

Answer : **13068**

Example 8 : 133 x 112

$$100$$

$$133 + 33$$
$$112 + 12$$

$$148 /_3 96$$

Answer : **14896**

When we observe carefully we find that we can do these types of sums mentally in the following way :-

L.H.S. : (First number PLUS the complement of the second number)
R.H.S. : (Product of the two complements)

Example : 104 × 106 = (104 + 06) / (04 × 06)
$$= 110 / 24$$
$$= 11024$$

1021 ×1004 = (1021 + 04) / (21 × 4)
$$= 1025 / 084$$
$$= 1025084$$

When we compare both the methods we find that the Vedic method makes our calculations very short.
112 x 123

Usual method	Vedic method
112	112 – 12
x 123	123 – 23
-------------	-------------
336	135 / ₂76
2240	
11200	= 13776

13776	

Exercise 4 .2 :

Solve the following sums mentally :-

(1) 17 × 19

(2) 16 × 18

(3) 104 × 107

(4) 117 × 109

(5) 114 × 108

(6) 145 × 104

(7) 121 × 109

(8) 1021 × 1003

(9) 1421 × 1002

(10) 1004 × 1042

Wavy Square

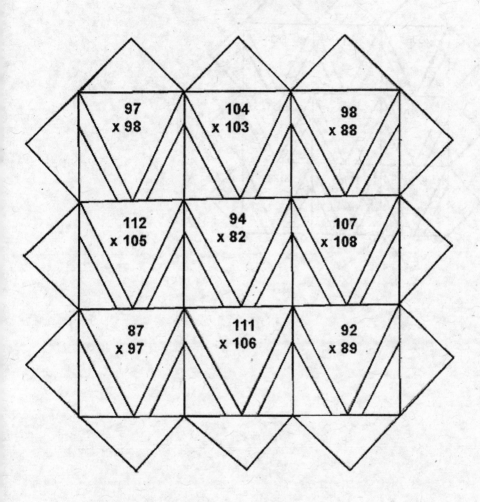

97
x 98

104
x 103

98
x 88

112
x 105

94
x 82

107
x 108

87
x 97

111
x 106

92
x 89

Solve the sums
If the answer is :   colour the shape :
less than 1000   yellow
more than 1000   orange

Answers to the Wavy Square

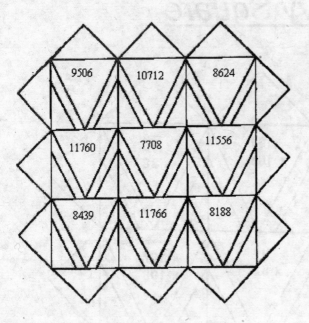

Colourful Pyramid

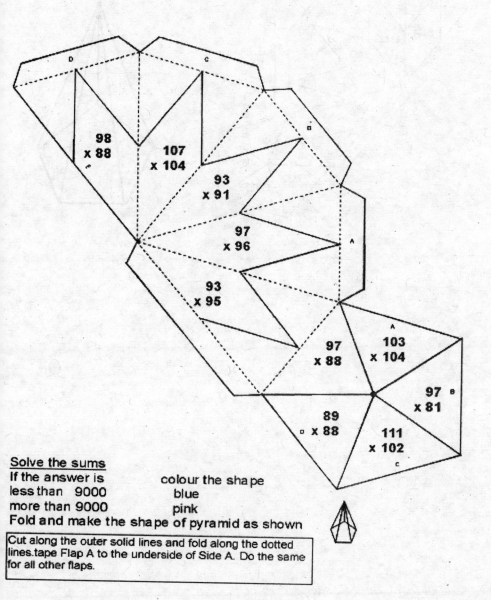

98 x 88

107 x 104

93 x 91

97 x 96

93 x 95

97 x 88

103 x 104

97 x 81

89 x 88

111 x 102

Solve the sums

If the answer is   colour the shape
less than   9000   blue
more than 9000   pink
Fold and make the shape of pyramid as shown

Cut along the outer solid lines and fold along the dotted lines.tape Flap A to the underside of Side A. Do the same for all other flaps.

Answers to Colourful Pyramid

Final Shape

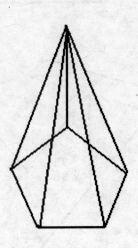

8624

11128

8463

9312

8835

8536

10712

7857

7832

11322

Multiplication of One more and One less than the Base number

Example 1 : 98 × 112

Step 1:

```
      100
----------------
  98   - 02
  112  + 12
----------------
```

Here both the numbers are close to the base 100, so we write our numbers as shown.

We place their complements next to them.

98 is 2 less than the base, so its complement is – 02 (minus 02).
112 is 12 more than the base, so its complement is +12 (positive 02).

Step 2 :

```
      100
----------------
  98      - 02
  112     + 12
----------------
  110  /   24
----------------
```

R.H.S. :
$$(- 02) \times (+ 12) = - 24$$
Since we get a negative number we write it as

$\overline{24}$

L.H.S. :
98 +12 = 110
or 112 - 2 = 110

Step 3 :

```
      100
----------------
  98   - 02
  112  +12
----------------
  109  /  76
----------------
```

Since we have a negative number in the R.H.S we have to make it positive as one part of the answer cannot be negative.
We take its complement using "All from 9 last from 10" and make it positive.

R.H.S. : complement of 24 = 76

We then subtract 1 extra (for coming out of complements) but from the L.H.S.

L.H.S. :   110 – 1 = 109

Answer : 98 × 112 = 10976

Example 2 : 107 × 93

Step 1:

         100
    ----
    107 + 07
     93 - 07

Here both the numbers are close to the base 100, so we write our numbers as shown.

We place their complements next to them.

107 is 7 more than the base, complement is + 07
93 is 7 less than the base, complement is - 07

Step 2:

         100
    ----
    107 ⤬ + 07 |
     93 ⤬ - 07 ↓
    ----
      100 / $\overline{49}$

R.H.S. :
   $(- 7) \times (+ 7) = - 49$

Since we get a negative number we write it as

$\overline{49}$

L.H.S. :
   107 - 07 = 100
or 93 + 07 = 100

Step 3:

         100
    ----
    107   + 07
     93   - 07
    ----
     99 / 51

Since we have a negative number in the R.H.S. we have to make it positive.
We take its complement using "All from 9 last from 10" to make it positive.

R.H.S. : The complement of 49 = 51

We then subtract 1 extra (for coming out of complements) but from the L.H.S.

L.H.S. : 100 – 1 = 99

Answer : 107 × 93 = 9951

Example 3 : 1033 × 997

Step 1:

```
        1000
---------------------
  1033   + 033
   997   - 003
---------------------
```

Here the base closest to the given numbers is 1000, so we write our numbers as shown.

We place their complements next to them.

1033 is 33 more than the base, complement is +033

997 is 3 less than the base, complement is - 003

Step 2:

```
         1000
---------------------
  1033  ⤬  + 033
   997     - 003
---------------------
  1030  /   ‾0‾9‾9‾
---------------------
```

R.H.S. :

$$(+33) × (- 3) = - 099$$

1000 is the base so we need 3 digits and since we get a negative number we write it as

$$\overline{099}$$

L.H.S. :
       1033 - 003 = 1030
or   997 + 033 = 1030

Step 3:

```
        1000
---------------------
  1033   + 033
   997   - 003
---------------------
  1029  /   901
---------------------
```

Since we have a negative number in the R.H.S. we have to make it positive.

R.H.S. : The complement of 099 = 901

We then subtract 1 extra (for coming out of complements) but from the L.H.S.

L.H.S. : 1030 – 1 = 1029

Answer : 1033 × 997 = **1029901**

Example 4 : 117 ×94

Step 1:

     100

  ----

117 + 17
94 - 06

Here both the numbers are close to the base 100, so we write our numbers as shown.

We write their complements next to them.

Step 2 :

       100

  ----

117   + 17
94   - 06

----

111 / $\overline{1\,02}$

-------- ∿ -------

R.H.S. :

     $(-17) \times (+6) = -102$

100 is the base so only 2 digits are required in the left side.

Since we get a negative number we write it as

$$\overline{}$$
$$_102$$

L.H.S. :

    117 - 06 = 111

or   94 + 17 = 111

Step 3 :

    100

  ----

117   +17
94   - 06

----

110 / $\overline{02}$

Before converting the negative R.H.S. to positive we *carry over the negative* 1 from right side to left side so that we have only 2 digits here as the base is 100.

L.H.S. :   111 – 1 = 110

Step 4:

    100

  ----

117 + 17
94 - 06

----

109 / 98

Now we will convert the negative number to positive.

R.H.S. : The complement of 02 = 98

We then subtract 1 extra (for coming out of complements) from the L.H.S.

L.H.S. : 110 – 1 = 109

Answer : 117 × 94 = **10998**

In the above example we have subtracted 1 twice from the L.H.S., once for a negative carry over and second time for coming out of complements.

Rules :

(1) We proceed our working in the same way as with any other base multiplication sum.

(2) Here we have a negative R.H.S. and to make it positive we take its complement. BUT before that we must make the number of digits in the R.H.S. equal to the number of zeros in the base, by either adding zeros or carrying over the negative carry over to the L.H.S.

(3) Then we take the complement of the number in the R.H.S. to make it positive.

(4) Subtract one extra from the L.H.S. for coming out of the complements.

Given below are a few more examples solving the sums in straight steps :

Example 5 : 96 x 107

$$100$$
$$\text{-----------}$$
$$96 \quad - 04$$
$$107 \quad + 07$$
$$\text{-----------}$$
$$103 \ / \ \overline{28}$$

$$= 102 \ / \ 72$$

Answer : **10272**

Example 6 : 76 x 112

$$100$$
$$\text{-----------}$$
$$76 \quad - 24$$
$$112 \quad +12$$
$$\text{-----------}$$
$$88 \ / \ _2 88$$

$$= 86 \ / \ \overline{88} \qquad = 85 \ / \ 12$$

Answer : **8512**

Example 7 : 126 x 94   **Example 8** : 1021 x 996

$$100$$

$$\overline{\quad\quad\quad}$$

$$\begin{array}{c} 126 + 26 \\ 94 - 06 \end{array}$$

$$\overline{\quad\quad\quad}$$

$$120 \;/\; \overline{1}56$$

$$= 119 \;/\; \overline{56} \;\; = 118 \;/\; 44$$

Answer : **11844**

$$1000$$

$$\overline{\quad\quad\quad\quad}$$

$$\begin{array}{c} 1021 + 021 \\ 996 - 004 \end{array}$$

$$\overline{\quad\quad\quad\quad}$$

$$1017 \;/\; \overline{084}$$

$$= 1016 \;/\; 916$$

Answer : **1016916**

These types of sums can also be done mentally but with a little care:

<u>L.H.S. :</u> (Number less than the base plus the complement of the number more than the base OR number more than base minus complement of the number less than the base)

<u>R.H.S. :</u> (Product of the two complements)

Example : 104×96 $= (104 - 04) \;/\; (04 \times -04)$

$$= 100 \;/\; -16$$

$$= 99 \;/\; 84$$

$$= 9984$$

994×1003 $= (994 + 003) \;/\; (- 006 \times 003)$

$$= 997 \;/\; - 018$$

$$= 996 / \; 982$$

$$= 996982$$

If we compare both the methods we see the Vedic method makes our calculations very short.

98 x 137

<table>
<tr><td>Usual method</td><td>Vedic method</td></tr>
</table>

Usual method

    98
  x 137

   686
  2940
  9800

  13426

Vedic method

  98   - 02
 137   + 37

 135/ 74
= 13426

Exercise 4 .3 :

Solve the following sums :-

(1) 12 × 8

(2) 96 × 104

(3) 127 x 96

(4) 103 × 67

(5) 82 × 114

(6) 1026 × 997

(7) 1009 × 982

(8) 976 × 1010

(9) 986 × 1005

(10) 985 × 1071

Answers

Exercise 4 .1

(1) 9702	(2) 9408	(3) 9024	(4) 8740
(5) 995004	(6) 985056	(7) 880236	(8) 99750100
(9) 99760144	(10) 98810888		

Exercise 4 .2

(1) 323	(2) 288	(3) 11128	(4) 12753
(5) 12312	(6) 15080	(7) 13189	(8) 1024063
(9) 1423842	(10) 1046168		

Exercise 4 .3

(1) 96	(2) 9984	(3) 12192	(4) 6901
(5) 9348	(6) 1022922	(7) 990838	(8) 985760
(9) 990930	(10) 1054935		

Chapter 5

WORKING BASE MULTIPLICATION

In the previous chapter we discussed multiplication of numbers close to any one base but when both the numbers are not close to a convenient power of 10 but are close to a multiple of the base. We take a multiple or sub multiple of a suitable base like 10, 100, 1000 etc, as our "**Working base**" and then do our multiplication taking the working base into account. We use the sutra

आनुरूप्येण

"Ānurūpyeṇa"

"Proportionately"

For example, if we have to multiply 46 x 53, here both our numbers are close to 50, taking 10 or 100 as the base is not convenient so we take a proportional base, 50 as our working base. It can be dealt with in two ways:

10 x 5 = 50, here **10** is the *theoretical base* and **50** is the *working base*.
100/2 = 50, here **100** is the *theoretical base* and **50** is the *working base*.

In both the above examples 50 is the working base but in both the cases the theoretical bases are different, the theoretical bases will make their impact when we have to decide the number of digits in the right hand side of the answer. For the theoretical base 10 we should have only one digit in the R.H.S. but for the theoretical base 100 we should have two digits in the R.H.S.

Let us take a few examples and understand it better.

<u>**Example 1**</u> : 52 × 54

Here both the numbers are far from both the base 10 and 100 so we take a working base of 50 as both the numbers are close to 50.

Step 1:

$$10 \times 5 = 50$$
----
54   +4
52   +2

We take 10 as the theoretical base and $10 \times 5 = 50$ as their working base.

We place their complements from the working base next to them.
54 is 4 more than the base 50
52 is 2 more than the base 50
We will get our answer in two parts.

Step 2 :

$$10 \times 5 = 50$$
----
54   + 4 | multiply
52   + 2 ↓

      / 8

R.H.S. :
$$(+ 4) \times (+ 2) = 8$$
Since 10 is the theoretical base so we need only 1 digit.

Step 3 :

$$10 \times 5 = 5\,0$$
----
54   + 4   cross
52   + 2   addition
----
56   /   8

L.H.S. :
We can get our answer in 4 different ways but we use either of the two methods that are the easiest.

$$54 + 2 = 56$$
or $$52 + 4 = 56$$

Step 4 :

$$10 \times 5 = 5\,0$$
----
54   + 4
52   + 2
----
280   /   8

As our sutra says "proportionately," we proportionately multiply our L.H.S. of the answer as we have done to get the working base from the theoretical base.

We multiply our L.H.S. by 5 as done in the base $(10 \times 5 = 50)$

L.H.S. : $56 \times 5 = 280$

Answer : $54 \times 52 = 2808$

Example 2 : 54 x 52

There is another method also to do the same sum where we take the working base as 100/2 = 50.

Step 1:   We take 100 as the theoretical base and $100/2 = 50$ as their working base.

```
100 / 2 = 50
-----------------
 54    + 4
 52    + 2
```

We place their complements from the working base next to them.

We will get our answer in two parts.

Step 2 :

```
     100/ 2 = 50
  -------------------
  54  ⤫  + 4 |
  52       + 2 |
  -------------------       ↓
  56   /   08
  -------------------
```

R.H.S. :
     $(+ 4) \times (+ 2) = 08$

Since 100 is the theoretical base here so we need 2 digits.

L.H.S. :
     $54 + 2 = 56$
or $52 + 4 = 56$

Step 3 :

```
100 / 2 = 50
-----------------
 54   + 4
 52   + 2
-----------------
 28  /  08
-----------------
```

As our sutra says "proportionately," we proportionately multiply our L.H.S. of the answer as we have done we have done to get the working base from the theoretical base.

Here we divide our L.H.S. by 2 as done in the base (100/2 = 50)

     $56/2 = 28$

Answer : $54 \times 52 = 2808$

We can choose any of the methods to solve our sum but we must follow the rule given below :

We multiply or divide our L.H.S. of the answer proportionately i.e. in the same proportion, as used to get the working base from the theoretical base.

Example 3 : 62×68

Step 1:

$10 \times 6 = 60$	

62	+2
68	+8

We take 10 as the theoretical base and $10 \times 6 = 60$ as their working base.
We place their complements from the working base next to them.
We will get our answer in two parts.

Step 2 :

$10 \times 6 = 60$

----

62 $\times$ + 2
68 $\times$ + 8

----

70 / $_1$6

------ ∿-------

R.H.S. :
$(+2) \times (+8) = {}_16$
Since 10 is the theoretical base here so we need 1 digit, we write 6 and *carry over* 1 to L.H.S.

L.H.S. :
$62 + 8 = 70$
OR $68 + 2 = 70$

Step 3 :

$10 \times 6 = 60$

----

62   + 2
68   + 8

----

421 / 6

As our sutra says "proportionately,"
we proportionately multiply our L.H.S. as we have done in the base.

$70 \times 6 = 420$
$420 + 1\ (carry) = 421$

We first proportionately multiply our L.H.S. then we add the carry over.

Answer : $62 \times 68 = \mathbf{4216}$

Example 4 : 62×68

Let's do the same example with a different working base.

Step 1:

$10 \times 7 = 70$	

62	- 8
68	- 2

We take 10 as the theoretical base and $10 \times 7 = 70$ as their working base.
We place their complements from the working base next to them.

We will get our answer in two parts.

Step 2 :

$$10 \times 7 = 70$$

| 62 | | - 8 | |
| 68 | | - 2 | |

----

  60   / ₁6

R.H.S. :

$$(-8) \times (-2) = 16$$

Since 10 is the theoretical base here, we need 1 digit.

We write **6** and *carry over* 1 to L.H.S.

L.H.S. :

    62 - 2 = 60

or   68 - 8 = 60

Step 3 :

$$10 \times 7 = 70$$

| 62 | - 8 |
| 68 | - 2 |

----

421 / 6

As our sutra says "proportionately," we proportionately multiply our L.H.S. as we have done to get the working base.

$$60 \times 7 = 420$$

L.H.S. :

$$420 + 1 \,(carry) = 421$$

Answer : $62 \times 68 = 4216$

Example 5 : 523×488

Step 1:

$$100 \times 5 = 500$$

| 523 | + 23 |
| 488 | - 12 |

We take 100 as the theoretical base and $100 \times 5 = 500$ as the working base.

We place their complements from the working base next to them.

We will get our answer in two parts.

Step 2 :

$$100 \times 5 = 500$$

| 523 | | + 23 | |
| 488 | | - 12 | |

----

511 / $\overline{2}$76

R.H.S. :

$$(+23) \times (-12) = -276 = \overline{276}$$

Since 100 is the theoretical base here so we need 2 digits, we write **76** and *carry over* *negative* 2 to L.H.S.

L.H.S. :

    523 - 12 = 511

or   488 + 12 = 511

Step 3 :

$100 \times 5 = 500$

523	+ 23
488	- 12

2553 / $\overline{76}$

As our sutra says "proportionately,"
we proportionately multiply our L.H.S. as we
have done to get the working base.

$511 \times 5 = 2555$

$2555 - 2\ (carry) = 2553$

We have a negative number in R.H.S. so a
negative carry over.

Step 4 :

$100 \times 5 = 500$

523	+ 23
488	- 12

2552 / 24

Since we still have a negative number in R.H.S.
and to make it positive we take its complement.

R.H.S. : The complement of 76 = 24
We subtract 1 extra from L.H.S. for coming out
of complements.

L.H.S. :
    $2553 - 1 = 2552$

Answer : $523 \times 488 = 255224$

Rules :

(1) We take a theoretical base which is common to both the numbers and then take a proportional base to the theoretical base as our working base.

(2) We work in the same way as base multiplication and get the two parts of the answer.

(3) We proportionally multiply or divide the L.H.S. of our answer as done to get the working base from the theoretical base.

(4) If we have a positive or a negative carry over of the R.H.S. depending on the theoretical base. We carry it over to the L.H.S. after proportionally multiplying it.

(5) If we have a negative R.H.S. we take its complement and then subtract 1 extra from the L.H.S. for coming out of complements.

Given below are a few more examples solving the sums in straight steps:

Example 6 : 19 × 499

$$100 \times 5 = 500$$

$$19 \quad - 481$$
$$499 \quad - 001$$

$$18 \quad / \quad _4 81$$
$$\times 5$$

$$90 \quad / \ 81$$
$$+4$$

$$94 \quad / \ 81$$

Answer : **9481**

Example 7 : 128 × 672

$$100 \times 7 = 700$$

$$128 \quad - 572$$
$$672 \quad - 028$$

$$100 \quad / \quad _{160} 16$$
$$\times 7$$

$$700 \quad / \ 16$$
$$+160$$

$$860 \quad / \ 16$$

Answer : **86016**

Example 8 : 231 × 582

$$100 \times 6 = 600$$

$$231 \quad - 369$$
$$582 \quad - 018$$

$$213 \quad / \quad _{66} 42$$
$$\times 6$$

$$1278 \quad / \ 42$$
$$+66$$

$$1344 \quad / \ 42$$

Answer : **134442**

Example 9 : 532 × 472

$$100 \times 5 = 500$$

$$532 \quad + 32$$
$$472 \quad - 028$$

$$504 \quad / \quad _8 96$$
$$\times 5$$

$$2520 \ / \ 96$$
$$-8$$

$$2512 \ / \ \overline{96}$$

$$2511 \ / \ 04$$

Answer : **251104**

The above examples are to show the working of triple digit numbers using working bases but they are more easily solved using the "*Urdhva triyagbhayam*" sutra method which is dealt with in the next chapter.

When we compare both the methods we find that the Vedic method makes our calculations very short.

524 x 512

<u>Usual method</u>

```
        524
      x 512
    -----------
       1048
       5240
     262000
    -----------
     268288
    -----------
```

<u>Vedic method</u>

```
    524   + 24
    512   + 12
    --------------
    536  /  ₂88
     x 5
    --------------
    2680 /  ₂88
  = 268288
```

Exercise 5 :

Solve the following sums :-

(1) 24 × 26

(2) 49 × 48

(3) 56 ×54

(4) 54 × 46

(5) 399 × 396

(6) 687 × 698

(7) 532 × 528

(8) 478 × 528

(9) 478 ×399

(10) 775 × 792

Answers

Exercise 5			
(1) 624	(2) 2352	(3) 3024	(4) 2484
(5) 158004	(6) 479526	(7) 280896	(8) 2523
(9) 109722	(10) 613800		

Zee Pattern

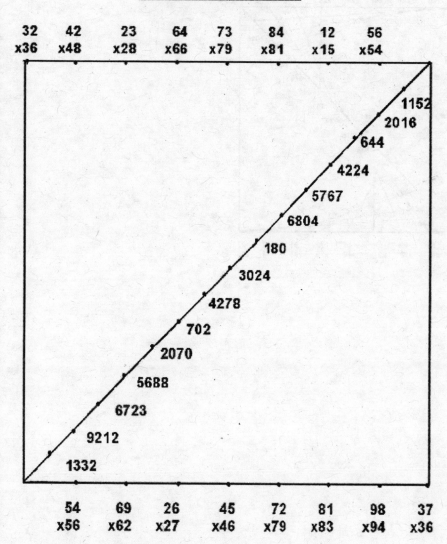

| 32 | 42 | 23 | 64 | 73 | 84 | 12 | 56 |
| x36 | x48 | x28 | x66 | x79 | x81 | x15 | x54 |

1152
2016
644
4224
5767
6804
180
3024
4278
702
2070
5688
6723
9212
1332

| 54 | 69 | 26 | 45 | 72 | 81 | 98 | 37 |
| x56 | x62 | x27 | x46 | x79 | x83 | x94 | x36 |

Solve the sums and join the dots
to the dots of their answers

Answer to Zee Pattern

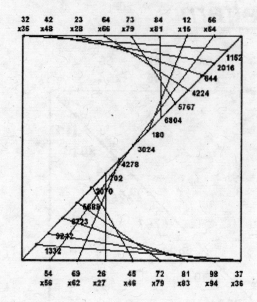

Chapter 6

MULTIPLICATION

In the previous chapter we have discussed specific cases of multiplication now we will understand the general formula of multiplication which will be applicable to multiplication of any number. We will use the sutra

ऊर्ध्वतिर्यग्भ्याम्

"Ūrdhvatiryagbhyāṃ"

"Vertical and Crosswise"

We will start with multiplication of a 2-digit number by another 2-digit number. In this type we get the answer in 3 steps which are explained below with the help of dot diagrams. Each dot represents a digit in the number and the line joining represents the digits to be multiplied.

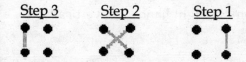

Step 1 : We start our multiplication from the right hand side. We vertically multiply the two digits in the right column.

Step 2 : We cross multiply the first digit of the first column with the second digit of the second column, and the second digit of the first column with the first digit of the second column, and add the two products.

Step 3 : We again vertically multiply the digits in the left column alone.

This is further made clear with the help of a few examples :

Example 1 : 21 × 32

Step 1:

```
  2 1  | vertical
x 3 2  |
       ↓
------
    2
------
```

Starting from the right hand side, vertically multiply

$1 × 2 = 2$

Step 2:

```
  2 ⟍⟋ 1  cross
x 3 ⟋⟍ 2
-----------
  7     2
-----------
```

Cross multiplication

Cross multiply $2 × 2 = 4$
Cross multiply $1 × 3 = 3$
Add both the products $4 + 3 = 7$

Step 3:

```
vertical | 2 1
       x | 3 2
         ↓
------
   6 7 2
------
```

Vertically multiply the digits in the left column.

$2 × 3 = 6$

Answer : $21 × 32 = 672$

Example 2 : 42 × 13

Step 1:

```
  4 2  | vertical
x 1 3  |
       ↓
------
      6
------
```

Starting from the right hand side, vertically multiply

$2 × 3 = 6$

Step 2:

```
  4 ⟍⟋ 2
x 1 ⟋⟍ 3
-----------
  ₁4   6
```

Cross multiplication

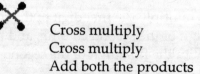

Cross multiply $3 × 4 = 12$
Cross multiply $1 × 2 = 2$
Add both the products $12 + 2 = {}_14$

We put down 4 and *carry over* 1.

Step 3:

Vertically multiply the digits in the left column

vertical | 4 2
    x | 1 3
----
    5 4 6

$4 \times 1 = 4$

$4 + 1(carry) = 5$

Answer : $42 \times 13 = $ **546**

Example 3 : 72×56

Step 1:

    7 2 | vertical
  x 5 6 |

      ₁2
--⤸----

Starting from the right hand side,

Vertically multiply

$2 \times 6 = {}_1 2$

We write down **2** and *carry over* **1** to next step,

Step 2:

    7   2 cross
  x 5   6

      ₅3 2
--⤸---------

Cross multiplication

Cross multiply $7 \times 6 = 42$
Cross multiply $5 \times 2 = 10$
Add both the products $42 + 10 = 52$

$52 + 1 \ (carry) = {}_5 3$

We put down **3** and *carry over* **5**

Step 3:

vertical | 7 2
    x | 5 6
----
   40 3 2

Vertically multiply

$7 \times 5 = 35$

$35 + 5(carry) = 40$

Answer : $72 \times 56 = $ **4032**

Example 4 : 37 × 89

Step 1:

$$\begin{array}{r} 3\ 7 \\ \times\ 8\ 9 \\ \hline {}_6 3 \end{array}$$ vertical

Starting from the right hand side

Vertically multiply
$$7 \times 9 = {}_6 3$$
We write down 3 and *carry over* 6 to next step.

Step 2:

$$\begin{array}{r} 3 \quad 7 \\ \times\ 8 \quad 9 \\ \hline {}_8 9\ 3 \end{array}$$ cross

Cross multiply $3 \times 9 = 27$
Cross multiply $8 \times 7 = 56$
Add both the products $27 + 56 = 83$

$$83 + 6\ (carry) = {}_8 9$$
We put down 9 and *carry over* 8

Step 3:

vertical $\begin{array}{r} 3\ 7 \\ \times\ 8\ 9 \\ \hline 3\ 2\ 9\ 3 \end{array}$

Vertically multiply

$$3 \times 8 = 24$$
$$24 + 8\ (carry) = 32$$

Answer : $37 \times 89 = 3293$

Example 5 : 46 × 27

$$\begin{array}{c} 4 \quad 6 \\ \times\ 2 \quad 7 \\ \hline 12\ {}_4 4\ {}_4 2 \end{array}$$

(1) $6 \times 7 = {}_4 2$
(2) $(4 \times 7) + (6 \times 2) = 40$
      $40 + 4 = {}_4 4$
(3) $4 \times 2 = 8$
      $8 + 4 = 12$

Answer : $46 \times 27 = \mathbf{1242}$

Example 6 : 57 × 83

$$
\begin{array}{r}
5\ 7 \\
\times\ 8\ 3 \\
\hline
47\ _73\ _21 \\
\hline
\end{array}
$$

(1) $7 \times 3 = {}_2 1$

(2) $(5 \times 3) + (7 \times 8) = 71$

 $71 + 2 = {}_7 3$

(3) $5 \times 8 = 40$

 $40 + 7 = 47$

Answer : 57 × 83 = **4731**

Rules :

(1) We vertically multiply the digits of the first column on the right.

(2) We then cross-multiply and add both the products.

(3) We then vertically multiply the digits of the column on the left.

(4) In each step we write down only one digit and carry over the other digits, if any, to the next step.

Given below are a few more examples solving the sums in straight steps :

Example 7 : 73 × 38

$$
\begin{array}{r}
7\ 3 \\
\times 3\ 8 \\
\hline
27\ _67\ _24 \\
\hline
\end{array}
$$

Answer : **2774**

Example 8 : 47 × 63

$$
\begin{array}{r}
4\ 7 \\
\times 6\ 3 \\
\hline
29\ _56\ _21 \\
\hline
\end{array}
$$

Answer : **2961**

Example 9 : 76 × 48

$$
\begin{array}{r}
7\ 6 \\
\times\ 4\ 8 \\
\hline
36\ _84\ _48 \\
\hline
\end{array}
$$

Answer : **3648**

Example 10 : 59 × 34

$$
\begin{array}{r}
5\ 9 \\
\times\ 3\ 4 \\
\hline
20\ _50\ _36 \\
\hline
\end{array}
$$

Answer : **2006**

When we compare both the methods we find that the Vedic method makes our calculations very short.

47 x 38

<u>Usual method</u>　　　　　　　　　　　　<u>Vedic method</u>

```
        47                                    47
      x 38                                  x 38
  ----------                            ----------
       376                               17 ₅8 ₅6
      1410                              ----------
  ----------                                 1786
      1786                              ----------
  ----------
```

<u>Exercise 6 .1 :</u>

Solve the following sums :-

(1) 24 ×56　　　　　　　　　(6) 62 × 81

(2) 43 × 29　　　　　　　　　(7) 19 × 25

(3) 76 × 24　　　　　　　　　(8) 15 × 79

(4) 13 × 43　　　　　　　　　(9) 32 × 65

(5) 23 × 75　　　　　　　　　(10) 71 × 29

Blooming Flower

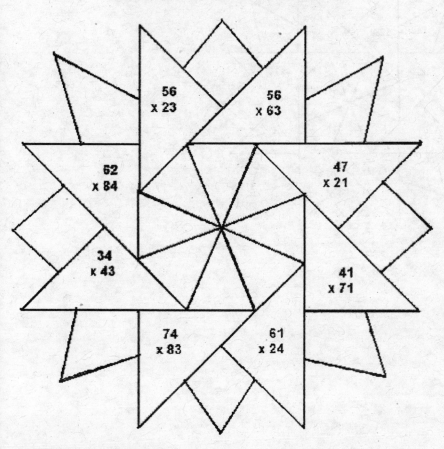

Solve the sums
If the answer is between :　　colour the shape :
0 - 2000　　yellow
more than 2000　　red

Answers to Blooming Flower

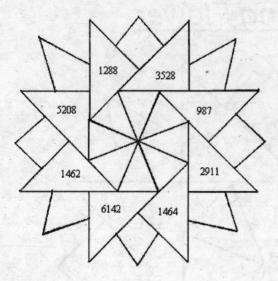

Trapezohedron

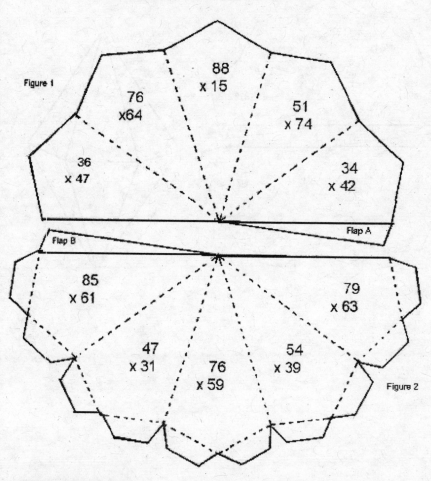

Figure 1

88
x 15

76
x64

51
x 74

36
x 47

34
x 42

Flap A

Flap B

85
x 61

79
x 63

47
x 31

76
x 59

54
x 39

Figure 2

Solve the sums
If the answer is :
Less than 3000   colour the shape :
green
more than 3000   red

Fold it and make it into the shape as shown

Cut out the pattern along the outer solid lines and fold along the
dotted lines. Tape flap A to the under side of figure 2. Then
carefully tape the other flap of figure 2 to the under side of figure 1.

Vedic Mathematics for all Ages

Answers to Trapezohedron

Final Shape

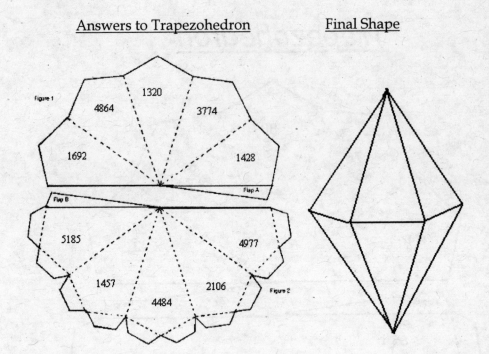

Three Digit Multiplication

The *"Urdhvatriyagbhayam"* sutra can be extended to the multiplication of 3 digit numbers by another 3 digit or 2 digit numbers also. The dot diagram explains the steps.

Step 5   Step 4   Step 3   Step 2   Step 1

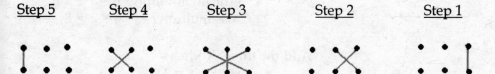

Each dot represents a digit in the number and the line represents the digits to be multiplied in each step. We start from the right side and move towards our left. At first we add one column one column more at a time in multiplication steps. And when we have taken all the columns then we start leaving one column from the right side at a time, till we are left with one column on the left.

Let us understand it with a few examples:

Example 1 : 123 × 645

Step 1:

```
  1 2 3
× 6 4 5
-----------
        15
------ ⤳ ----
```

Starting from the right hand side

Vertically multiply
$3 \times 5 = {}_1 5$

We write 5 and *carry over* 1 to next step.

Step 2:

```
 1 2   3
× 6 4   5
-----------------
      23  5
-- ⤳ --------
```

We cross multiply digits of two columns from the right.

Cross multiply $2 \times 5 = 10$
Cross multiply $4 \times 3 = 12$

Add both the products
$10 + 12 = 22$
$22 + 1 \ (carry) = {}_2 3$
We put down 3 and *carry over* 2

Step 3:

$$_3 3 \quad 3 \quad 5$$

We now Cross and Vertically multiply digits of all the three columns. All the digits are used in this step.

Cross multiply $1 \times 5 = 5$
Vertically multiply $6 \times 3 = 18$
Cross multiply $4 \times 2 = 8$

Add the three products
$$5 + 18 + 8 = 3\,1$$
$$31 + 2 \,(\textit{carry}) = {}_3 3$$
We put down 3 and *carry over* 3

Step 4:

$$\begin{array}{ccc} 1 & 2 & 3 \\ \times\,6 & 4 & 5 \end{array}$$
$$_1 9 \quad 3 \quad 3 \quad 5$$

We now cross multiply the digits of the two columns from the left.

Cross multiply $1 \times 4 = 4$
Cross multiply $6 \times 2 = 12$

Add both the products
$$4 + 12 = 16$$
$$16 + 3 \,(\textit{carry}) = {}_1 9$$
We put down 9 and *carry over* 1

Step 5:

$$\begin{array}{ccc} & 1 & 2 & 3 \\ \times & 6 & 4 & 5 \end{array}$$
$$7 \cdot 9 \quad 3 \quad 3 \quad 5$$

We vertically multiply the digits of the column on the left side.

Vertically multiply
$$1 \times 6 = 6$$
$$6 + 1 \,(\textit{carry}) = 7$$

Answer : $123 \times 645 = 79335$

Example 2 : 241×324

Step 1:

$$\begin{array}{ccc} 2 & 4 & 1 \\ \times\ 3 & 2 & 4 \end{array}$$
$$ 4$$

Starting from the right hand side

Vertically multiply $1 \times 4 = 4$

Step 2:

```
    2 4       1
  × 3 2       4
  ---------------
         ₁8  4
  ---
```

Cross multiply

Cross multiply $4 \times 4 = 16$
Cross multiply $2 \times 1 = 2$

Add both the products $16 + 2 = {}_18$

We put down 8 **and** *carry over* 1

Step 3:

```
  2  | 4  1
  × 3 | 2  4
  ---------------
       ₂0  8  4
  ---
```

Cross and Vertically multiply

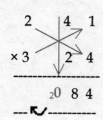

Cross multiply $2 \times 4 = 8$
Vertically multiply $3 \times 1 = 3$
Cross multiply $4 \times 2 = 8$

Add the three products
$$8 + 3 + 8 = 19$$
$$19 + 1 \ (carry) = {}_20$$
We put down 0 **and** *carry over* 2

Step 4:

```
   2    4  1
  × 3    2  4
  ---------------
    ₁8 0 8 4
```

Cross multiply

Cross multiply $2 \times 2 = 4$
Cross multiply $3 \times 4 = 12$

Add both the products
$$4 + 12 = 16$$
$$16 + 2 \ (carry) = {}_18$$
We put down 8 **and** *carry over* 1

Step 5:

```
   | 2 4 1
  × | 3 2 4
   ---------
  7 8 0 8 4
  ---------
```

Vertically multiply

Vertically multiply
$2 \times 3 = 6$
$6 + 1 \ (carry) = 7$

Answer : $241 \times 324 = 78084$

Example 3 : 521 × 234

```
  5 2 1
×  2 3 4
----------------
12 ₂1 ₂9 ₁1 4
----------------
```

(1) $4 \times 1 = 4$

(2) $(4 \times 2) + (3 \times 1) = {}_11$

(3) $(5 \times 4) + (2 \times 1) + (2 \times 3) = 28$
      $28 + 1 \, (carry) = {}_29$

(4) $(5 \times 3) + (2 \times 2) = 19$
      $19 + 2 \, (carry) = {}_21$

(5) $5 \times 2 = 10$
      $10 + 2 \, (carry) = 12$

Answer : 521 x 234 = **121914**

Example 4 : 362 × 724

```
  3 6 2
×  7 2 4
----------------
26 ₅2 ₄0 ₂8 8
----------------
```

(1) $2 \times 4 = 8$

(2) $(6 \times 4) + (2 \times 2) = {}_28$

(3) $(3 \times 4) + (7 \times 2) + (6 \times 2) = 38$
      $38 + 2 \, (carry) = {}_40$

(4) $(3 \times 2) + (7 \times 6) = 48$
      $48 + 4 \, (carry) = {}_52$

(5) $3 \times 7 = 21$
      $21 + 5 \, (carry) = 26$

Answer : 362 x 754 = **262088**

Rules :

(1) We vertically multiply the digits of the first column on the right side.

(2) We take two columns from the right and cross multiply their digits and add the product.

(3) We add another column to our operations i.e. we take all the three columns. We cross multiply the digits of the first and third columns and vertically multiply the digits of the middle column.

(4) Now since we have taken all the columns in our operation we start leaving one column at a time from the right again.

(5) Now we cross multiply the digits of the two column of the left side.

(6) We again leave another column from the right and vertically multiply the digits of the left most column.

(7) In each step we write down only one digit and carry over the other digits if any to the next step.

Given below are a few more examples solving sums in straight steps :

Example 5 : 136 x 241

$$
\begin{array}{r}
1\ 3\ 6 \\
\times 2\ 4\ 1 \\
\hline
3\ _{1}2\ _{2}7\ _{2}7\ 6 \\
\hline
\end{array}
$$

Answer : **32776**

Example 6 : 362 x 913

$$
\begin{array}{r}
3\ 6\ 2 \\
\times\ 9\ 1\ 3 \\
\hline
33\ _{6}0\ _{3}5\ _{2}0\ 6 \\
\hline
\end{array}
$$

Answer : **330506**

Example 7 : 527 x 183

$$
\begin{array}{r}
5\ 2\ 7 \\
\times\ 1\ 8\ 3 \\
\hline
9\ _{6}6\ _{4}4\ _{4}4\ _{2}1 \\
\hline
\end{array}
$$

Answer : **96441**

Example 8 : 532 x 216

$$
\begin{array}{r}
5\ 3\ 2 \\
\times\ 2\ 1\ 6 \\
\hline
11\ _{4}4\ _{3}9\ _{2}1\ _{1}2 \\
\hline
\end{array}
$$

Answer : **114912**

When we compare both the methods we find that the Vedic method makes our calculations very short.

47 x 38

<u>Usual method</u>   <u>Vedic method</u>

     763   763
  x 348   x 348
 ----------   ----------------

  6104 $26\ _55\ _95\ _62\ _24$
30520   ----------------
228900   265524

265524

Till now we have seen multiplication of a three-digit number by another three-digit number, now let us apply the same method when a three-digit number is to be multiplied by a two-digit number.

Example 9 : 214 × 23

We follow the same steps for this type of sum also .We just convert our two digit number to a three·digit number by adding a zero before the number.

We write it as 214 × 023

    2 1 4   (1) $4 \times 3 = {}_12$
× 0 2 3   (2) $(3 \times 1) + (4 \times 2) = 11$
-------------- $11 + 1\ (carry) = {}_12$
4 9 2 2   (3) $(3 \times 2) + (4 \times 0) + (1 \times 2) = 8$
-------------- $8 + 1\ (carry) = 9$
                     (4) $(2 \times 2) + (0 \times 1) = 4$
                     (5) $2 \times 0 = 0$

Answer : 214 x 23 = 4922

Example 10 : 57 × 346

We write it as   057 × 346

$$
\begin{array}{r}
0\,5\,7 \\
\times\,3\,4\,6 \\
\hline
1\,9\,7\,2\,2 \\
\hline
\end{array}
$$

(1) $7 \times 6 = {}_4 2$
(2) $(5 \times 6) + (4 \times 7) = 58 + 4 = {}_6 2$
(3) $(0 \times 6) + (3 \times 7) + (5 \times 4) = 41$
    $41 + 6 = {}_4 7$
(4) $(0 \times 4) + (3 \times 5) = 15$
    $15 + 4 = {}_1 9$
(5) $0 \times 3 = 0$
    $0 + 1 = 1$
Answer : 57 x 346 = **19722**

Rules :
(1) If we have to multiply a three digit number by any two digit number we add zero before the number and make it a three digit number for practical purpose.

(2) We continue further as in the case of multiplication of a three-digit by three-digit number.

When we compare both the methods we find that the Vedic method makes our calculations very short.

736 x 834

Usual method	Vedic method

Usual method

$$
\begin{array}{r}
736 \\
\times\,834 \\
\hline
2944 \\
22080 \\
588800 \\
\hline
613824 \\
\hline
\end{array}
$$

Vedic method

$$
\begin{array}{r}
736 \\
\times\,834 \\
\hline
61\,{}_5 3\,{}_8 8\,{}_3 2\,{}_2 4 \\
\hline
613824 \\
\end{array}
$$

Exercise 6 .2 :

Solve the following sums :-

(1) 123 × 215

(2) 236 × 213

(3) 721 × 72

(4) 281 × 19

(5) 982 × 124

(6) 724 × 362

(7) 289 × 727

(8) 473 × 67

(9) 926 × 94

(10) 836 × 492

Double Pentagonal Pyramid

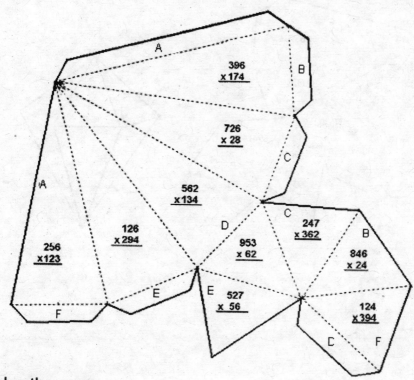

A

396
x 174

B

726
x 28

C

A

562
x134

126
x 294

D

953
x 62

C

247
x 362

B

846
x 24

256
x123

256
x123

E

E

527
x 56

124
x394

F

C

F

Solve the sums
If the answer is between :
0 and 35000
35001 and 70000
more than 70000

Colour the shape
red
blue
green

Cut out the pattern along the outer solid lines and fold along the dotted lines. Tape Flap A to the underside of side A. Do the same for all other flaps. Fold and make it into the shape as shown.

Answers to Double Pentagonal Pyramid

Final Shape

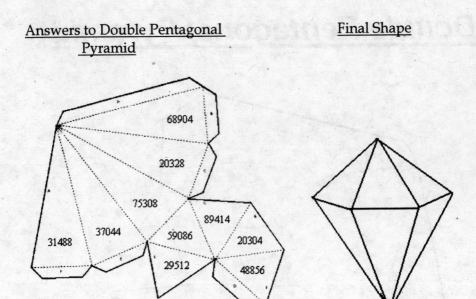

Multiplication of a Four digit Number by another Four digit Number

The procedure of multiplication of two 4-digit numbers remains the same only the pattern is extended. We always start multiplying digits in the right most column and in each next step we keep adding a column to our operation. When we have reached a step where we have taken all the columns in our multiplication we start leaving columns again from the right and in the last step we just have a single left most column. And in each step we either do a vertical or cross wise multiplication or both depending on the number of columns we have in that step.

A four digit into a four digit multiplication has 7 steps. The dot diagram shows the steps.

Step 7	Step 6	Step 5	Step 4	Step 3	Step 2	Step 1

Example 1 : 3124 × 2015

```
    3 1 2 4
  × 2 0 1 5
  ---------------
  6 2 9 4 8 6 0
  ---------------
```

(1) $4 \times 5 = {}_2 0$

(2) $(2 \times 5) + (1 \times 4) = 14$

  $14 + 2\ (carry) = {}_1 6$

(3) $(1 \times 5) + (0 \times 4) + (2 \times 1) = 7$

  $7 + 1\ (carry) = 8$

(4) $(3 \times 5) + (2 \times 4) + (1 \times 1) + (0 \times 2) = {}_2 4$

(5) $(3 \times 1) + (2 \times 2) + (1 \times 0) = 7$

  $7 + 2\ (carry) = 9$

(6) $(3 \times 0) + (2 \times 1) = 2$

(7) $3 \times 2 = 6$

Answer : 3124 x 2015 = **6294860**

Example 2 : 2436 × 372

Here we have one four digit and one three-digit number, so for practical purpose we convert both our numbers as four digit numbers and start our calculations. We write it as 2436 x 0372

$$\begin{array}{r} 2\ 4\ 3\ 6 \\ \times\ 0\ 3\ 7\ 2 \\ \hline 9\ 0\ 6\ 1\ 9\ 2 \\ \hline \end{array}$$

(1) $6 \times 2 = {}_1 2$
(2) $(3 \times 2) + (7 \times 6) = 48$
      $48 + {}_1 (carry) = {}_4 9$
(3) $(4 \times 2) + (3 \times 6) + (3 \times 7) = 47$
      $47 + {}_4 (carry) = {}_5 1$
(4) $(2 \times 2) + (0 \times 6) + (4 \times 7) + (3 \times 3) = 41$
      $41 + {}_5 (carry) = {}_4 6$
(5) $(2 \times 7) + (0 \times 3) + (4 \times 3) = 26$
      $26 + {}_4 (carry) = {}_3 0$
(6) $(2 \times 3) + (0 \times 4) = 6$
      $6 + {}_3 (carry) = 9$
(7) $2 \times 0 = {}_0 0$

Answer : 2436 x 372 = **906192**

Example 3 : 522 × 2431

$$\begin{array}{r} 0\ 5\ 2\ 2 \\ \times\ 2\ 4\ 3\ 1 \\ \hline 1\ 2\ 6\ 8\ 9\ 8\ 2 \\ \hline \end{array}$$

(1) $2 \times 1 = {}_2$
(2) $(2 \times 1) + (3 \times 2) = {}_8$
(3) $(5 \times 1) + (4 \times 2) + (2 \times 3) = {}_1 9$
(4) $(0 \times 1) + (2 \times 2) + (5 \times 3) + (4 \times 2) = 27$
      $27 + {}_1 (carry) = {}_2 8$
(5) $(0 \times 3) + (2 \times 2) + (5 \times 4) = 24$
      $24 + {}_2 (carry) = {}_2 6$
(6) $(0 \times 4) + (2 \times 5) = 10$
      $10 + {}_2 (carry) = {}_1 2$
(7) $0 \times 2 = 0$
      $0 + {}_1 (carry) = 1$

Answer : 522 x 2431 = **1268982**

Example 4 : 4763 × 73

4 7 6 3
× 0 0 7 3
----
0 3 4 7 6 9 9
----

(1) $3 \times 3 = 9$
(2) $(6 \times 3) + (7 \times 3) = {}_3 9$
(3) $(7 \times 3) + (0 \times 3) + (6 \times 7) = 63$
    $63 + 3 \ (carry) = {}_6 6$
(4) $(4 \times 3) + (0 \times 3) + (7 \times 7) + (0 \times 6) = 61$
    $61 + 6 \ (carry) = {}_6 7$
(5) $(4 \times 7) + (0 \times 6) + (0 \times 7) = 28$
    $28 + 6 \ (carry) = {}_3 4$
(6) $(4 \times 0) + (0 \times 7) = 0$
    $0 + 3 \ (carry) = 3$
(7) $4 \times 0 = 0$

Answer : 4763 x 73 = **347699**

General Rules :

(1) We always start with the column from the right and in each next step we keep adding a column to our operation. When we have reached a step where we have taken all the columns in our sum we start leaving columns again from the right and in the last step we just have a single left most column.

(2) In each step we either do a vertical or cross wise multiplication or both depending on the number of columns we have in that step.

(3) If there is more than one column in our multiplication then we add all the products.

(4) In each step we write down only one digit and carry over the other digits, if any, to the next step.

Exercise 6 .3 :

Solve the following sums :-

(1) 1023 × 2132

(2) 3125 × 126

(3) 5216 × 29

(4) 27 × 3271

(5) 521 × 6215

(6) 5612 × 2135

(7) 752 × 1920

(8) 9210 × 28

(9) 617 × 5611

(10) 6403 × 45

Answers

Exercise 6 .1

(1) 1344	(2) 1247	(3) 1824	(4) 559
(5) 1725	(6) 5022	(7) 475	(8) 1185
(9) 2080	(10) 2059		

Exercise 6 .2

(1) 26445	(2) 50268	(3) 51912	(4)5339
(5)121768	(6) 262088	(7) 210103	(8) 31691
(9) 87044	(10) 411312		

Exercise 6 .3

(1) 2181036	(2) 393750	(3) 151264	(4) 88317
(5) 3238015	(6) 11981620	(7) 1443840	(8) 257880
(9) 3461987	(10) 288135		

Chapter 7

ALGEBRA

Now that we are well versed with multiplication of numbers using the sutra

ऊर्ध्वतिर्यग्भ्याम्

"Ūrdhvatiryagbhyām"

"Vertical and Crosswise"

the multiplication of algebraic equations becomes very easy, as it is based on the same sutra and we follow the same method.

We will begin with algebraic equations, which have two terms

<u>Example 1</u> : (a +2b) (3a +b)

<u>Step1:</u>

```
a  b
1  2
3  1
-----------
```

We write the two variables on top and their coefficients from each equation below them.

For (a +2b) we write 1 and 2 as the coefficients of 'a' and 'b' respectively.
Similarly for (3a + b) we write 3 and 1 as coefficients of 'a' and 'b' respectively.

<u>Step 2:</u>

```
a     b
1    2
3    1
-----------
3 + 7 + 2
```

Now we just do our vertical and crosswise multiplication of 2 digits starting from the right.

(1) $2 \times 1 = + 2$ (vertical)
(2) $(1 \times 1) + (3 \times 2) = 1 + 6 = + 7$ (crosswise)
(3) $1 \times 3 = + 3$ (vertical)

Step 3:

a	b
1	2
3	1

$3a^2 + 7ab + 2b^2$

Starting from the right, we add the variables to these coefficients.

1) b^2 to 2 (we had multiplied the coefficients of b in this step)

2) ab to 7 (we had cross multiplied coefficients of a and b in this step)

3) a^2 to 3 (we had multiplied the coefficient of a in this step)

Answer : $(a+2b)(3a+b) = 3a^2 + 7ab + 2b^2$

Example 2 : $(2a + 3b)(4a + 2b)$

Step 1:

a	b
2	3
4	2

We write the two variables on top and their coefficients from each equation below them.
For (2a +3b) we write 2 and 3 as coefficients of '*a*' and '*b*' respectively and for (4a +2b) we write 4 and 2 as coefficients of '*a*' and '*b*' respectively.

Step 2:

$8 + 16 + 6$

We now perform vertical and crosswise multiplication from the right.

(1) $3 \times 2 = {}^+6$

(2) $(2 \times 2) + (4 \times 3) = 4 + 12 = {}^+16$

Here we do not carry forward any digit of the double digit (16) to the next step as done in multiplication but take the whole number in the same step so we write 16 as it is.

(3) $2 \times 4 = {}^+8$

These are the coefficients of the answer.

Step 3:

a	b
2	3
4	2

$8a^2 + 16ab + 6b^2$

We add the variable to our coefficients starting from the right as b^2, ab, a^2 respectively.

(1) b^2 to 6 (we had multiplied the coefficients of b in this step)

(2) ab to 16 (we had cross multiplied coefficients of a, b in this step)

(3) a^2 to 8 (we had multiplied the coefficient of a in this step)

Answer : $(2a+3b)(4a+2b) = 8a^2 + 16ab + 6b^2$

Example 3 : $(7x + 3y)(2x + 4y)$

Step 1:

x	y
7	3
2	4

We put down the coefficient below the variable for both the equations.

Step 2:

x   y

14 + 34 + 12

Vertical and crosswise multiplication from the right.

(1) $4 \times 3 = +12$

(2) $(7 \times 4) + (2 \times 3) = 28 + 6 = +34$

(3) $7 \times 2 = +14$

Step 3:

x	y
7	3
2	4

$14x^2 + 34xy + 12y^2$

We add the variables to the coefficients of the answer here it will be y^2, xy, x^2 respectively starting from the term on the right.

Answer : $(7x+3y)(2x+4y) = 14x^2 + 34xy + 12y^2$

Example 4 : $(2x - 3y)(5x + 8y)$

Step 1:

x	y
2	- 3
5	8

We put down the coefficients along their respective signs below the variables. A negative term will have a negative coefficient (-3 for $-3y$ in 1st equation).

Step 2:

x   y

2   - 3
5   8

10 + 1 - 24

Vertical and crosswise multiplication from the right.

(1) $-3 \times 8 = -24$

(2) $(8 \times 2) + (5 \times -3) = 16 - 15 = +1$

(3) $5 \times 2 = +10$

The first part will have a negative number.

Step 3:

x	y
2	- 3
5	8

$10x^2 + xy - 24y^2$

We add the variables to the coefficients of the answer.

Answer : $(2x - 3y)(5x+8y) = 10x^2 + xy - 24y^2$

Example 5 : $(3x + 4y) (2x - 7y)$

Step 1:

x	y
3	4
2	-7

We put down the coefficients below the variables.

(- 7 for – 7y in 2nd equation)

Step 2:

x	y
3	4
2	-7

----

6 – 13 – 28

Vertical and crosswise multiplication from the right.

(1) $4 \times -7 = -28$

(2) $(3 \times -7) + (4 \times 2) = -21 + 8 = -13$

(3) $3 \times 2 = +6$

Here both the first and the second terms will be negative numbers.

Step 3:

x	y
3	4
2	-7

$6x^2 - 13xy - 28y^2$

We add the variables to the coefficients of the answer.

Answer : $(3x+4y) (2x - 7y) = 6x^2 - 13xy - 28y^2$

Example 6 : $(9x - 7y) (2x - 3y)$

x	y
9	-7
2	-3

$18x^2 – 41xy + 21y^2$

(1) $-7 \times -3 = +21$

(2) $(9 \times -3) + (-7 \times 2) = -27 - 14 = -41$

(3) $9 \times 2 = +18$

We then add the variables to the coefficients.

Answer : $(9x+7y) (2x - 3y) = 18x^2 - 41xy +21y^2$

Example 7: $(4x - 5y) (3x - 2y)$

x	y
4	-5
3	-2

$12x^2 – 23xy + 10y^2$

(1) $-5 \times -2 = +10$

(2) $(4 \times -2) + (3 \times -5) = -8 - 15 = -23$

(3) $4 \times 3 = +12$

We then add the variables to the coefficients.

Answer : $(4x -5y) (3x - 2y) = 12x^2 - 23xy +10y^2$

Rules :

(1) We first write our variables.

(2) We then write the coefficients of both the equations below them.

(3) We multiply the numbers by the vertical and crosswise multiplication.

(4) We don't take forward any carry over, but put down the whole number in that step only. These numbers form the coefficients of the answer.

(5) We then add the variables to the coefficients of the answer taking the variables into consideration.

Given below are a few more examples solving sums in straight steps :

Example 8 : $(4x - 3y)(2x + 3y)$

$$
\begin{array}{cc}
x & y \\
4 & -3 \\
2 & +3 \\
\hline
\end{array}
$$

$8x^2 + 6xy - 9y^2$

Answer : $8x^2 + 6xy - 9y^2$

Example 9 : $(2a + 5b)(4a - 2b)$

$$
\begin{array}{cc}
a & b \\
2 & +5 \\
4 & -2 \\
\hline
\end{array}
$$

$8a^2 + 16ab - 10b^2$

Answer : $8a^2 + 16ab - 10b^2$

Example 10 : $(7x - 3y)(4x - 8y)$

$$
\begin{array}{cc}
x & y \\
7 & -3 \\
4 & -8 \\
\hline
\end{array}
$$

$28x^2 - 68xy + 24y^2$

Answer : $28x^2 - 68xy + 24y^2$

Example 11 : $(x - 7y)(3x - 8y)$

$$
\begin{array}{cc}
x & y \\
1 & -7 \\
3 & -8 \\
\hline
\end{array}
$$

$3x^2 - 29xy + 56y^2$

Answer : $3x^2 - 29xy + 56y^2$

When we compare both the methods we find that the Vedic method makes our calculations very short.

(4a – 3b)(4a + 2b)

<u>Usual method</u>   <u>Vedic method</u>

(4a – 3b)(4a +2b)   a   b
=4a (4a +2b) - 3b (4a +2b)   4   - 3
=16a ^{2} + 8ab – 12ab – 6b²   4   + 2
=16a² – 4ab – 6b²   -------------------
                                                              16a² –4ab – 6b²

<u>Exercise 7 .1 :</u>

Solve the following algebraic equations :-

(1) (2a+3b) (3a+4b)   (6) (23x - 5y) (6x+2y)

(2) (7a-2b) (3a+5b)   (7) (11a – 2by) (3a - 9b)

(3) (4x+y) (2x+3y)   (8) (5x+2y) (3x-9y)

(4) (7x-2y) (5x+4y)   (9) (3p – 2q) (7p – 9q)

(5) (17x+3y) (2x-7y)   (10) (17m+4n) (5m – 7n)

Multiplication of Algebraic Equation with larger power of Variables

In sums where we have variables with different powers of x, we still follow the same *"Ūrdhvatiryagbhyām"* sutra i.e. vertical and crosswise multiplication. We must just take care of the different powers of our variable.

Example 1 : $(2x^2 + x + 3) (3x^2 + 2x + 4)$

Step 1:

x^2	x	-
2	1	3
3	2	4

Here our variables are x^2, x, and independent term, we write it as x^2, x, - . ('-' for the independent term) Here again we put down our coefficients below the variables for both the equations.

Step 2:

x^2	x	-
2	1	3
3	2	4

6+7+19+10+12		

Vertical and crosswise multiplication from the right.
(1) $(4 \times 3) = +12$
(2) $(4 \times 1) + (2 \times 3) = +10$
(3) $(2 \times 4) + (3 \times 3) + (1 \times 2) = +19$
(4) $(2 \times 2) + (1 \times 3) = +7$
(5) $(2 \times 3) = +6$

Step 3:

x^2	x	-
2	1	3
3	2	4

$6x^4 + 7x^3 + 19x^2 + 10x + 12$		

We now add the variables to our coefficients. Starting from the right we start with independent term, our next term will be x and then we keep increasing the power of x by 1 as we move towards the left. We add

(1) No variable 12 (multiplied the coefficients of 2 independent terms)
(2) x to 10 (multiplied x term to an independent term)
(3) x^2 to 19 (multiplied x^2 term to an independent term and two x terms)
(4) x^3 to 7 (multiplied x^2 term to x term)
(5) x^4 to 6 (multiplied two x^2 terms)

Answer : $(2x^2 + x + 3) (3x^2 + 2x + 4)$
$= 6x^4 + 7x^3 + 19x^2 + 10x + 12$

Example 2 : $(x^2 - x + 5) (2x^2 + 3x - 2)$

Step 1:

x^2	x	-
1	-1	+5
2	+3	-2

Here our variables are x^2, x, and independent term, We write it as x^2, x , - ,
('-' for the independent term)
We place our coefficients below the variables respectively for both the equations along with their signs.

Step 2:

x^2	x	-
1	-1	+5
2	+3	-2

$2 + 1 + 5 + 17 - 10$

Vertical and crosswise multiplication from the right.

(1) $(5 \times -2) = -10$
(2) $(-1 \times -2) + (5 \times 3) = 2 + 15 = +17$
(3) $(1 \times -2) + (2 \times 5) + (-1 \times 3) = -2 + 10 - 3 = +5$
(4) $(1 \times 3) + (-1 \times 2) = 3 - 2 = +1$
(5) $2 \times 1 = 2$

Step 3:

x^2	x	-
1	-1	+5
2	+3	-2

$2x^4 + x^3 + 5x^2 + 17x - 10$

We add the variables to the coefficients of our answer starting from the right we start independent term, our next term will be x and then we keep increasing the power of x by 1.
We add
(1) No variable to -10
(2) x to 17
(3) x^2 to 5
(4) x^3 to 1
(5) x^4 to 2

Answer : $(x^2 - x + 5) (2x^2 + 3x - 2)$
= $2x^4 + x^3 + 5x^2 + 17x - 10$

Example 3 : $(2x^2 - 3x - 5) (3x^2 + 2x - 3)$

x^2	x	-
2	-3	-5
3	+2	-3

$6 - 5 - 27 - 1 + 15$

We just perform vertical and crosswise multiplication and then add the variables to get our answer.

Answer : $(2x^2 - 3x - 5) (3x^2 + 2x - 3)$
= $6x^4 - 5x^3 - 27x^2 - x + 15$

Example 4 : $(7x^2 + 2x - 6)(3x^2 + 2x - 8)$

x^2	x	-
7	2	-6
3	2	-8

---

21 +20 - 70 - 28 +48

We just perform vertical and crosswise multiplication and then add the variables to get our answer.

Answer : $(7x^2 + 2x - 6)(3x^2 + 2x - 8)$
$$= 21x^4 + 20x^3 - 70x^2 - 28x + 48$$

Rules :

(1) We start by putting down the variables and then the coefficients of our equations below them.

(2) We perform vertical and crosswise multiplication for the coefficients.

(3) We then add the variables to the coefficients of the answer taking the variables into consideration.

Note:- Instead of placing '–' in place of the independent term we can also write it as x^0 as we know that any number or variable to the power of zero is equal to 1, similarly $x^0 = 1$ which will mean an independent term.

Given below are a few more examples solving sums in straight steps :

Example 5 : $(x^2 + x - 6)(3x^2 + 2x +3)$

x^2	x	-
1	1	- 6
3	2	3

$3x^4 +5x^3 - 13x^2 - 9x - 18$

Answer : $3x^4 + 5x^3 - 13x^2 - 9x - 18$

Example 6 : $(6x^2 + 3x - 6)(x^2 - 2x +2)$

x^2	x	-
6	3	- 6
1	-2	2

$6x^4 - 9x^3 +0x^2 +18x - 12$

Answer : $6x^4 - 9x^3 + 18x - 12$

When we compare both the methods we find that the Vedic method makes our calculations very short.

$(2x^2 + 5x - 6) (x^2 + 7x + 3)$

<table>
<tr><td>Usual method</td><td>Vedic method</td></tr>
</table>

Usual method	Vedic method

$(2x^2 + 5x - 6) (x^2 + 7x + 3)$

$= 2x^2 (x^2 + 7x + 3) + 5x (x^2 + 7x + 3)$

$- 6 (x^2 + 7x + 3)$

$= 4x^4 + 14x^3 + 6x^2 + 5x^3 + 35x^2$

$+ 15x - 6x^2 - 42x - 18$

$= 2x^4 + 19x^3 + 35x^2 - 27x - 18$

x^2	x	-
2	5	- 6
1	7	3

$2x^4 + 19x^3 + 35x^2 - 27x - 18$

Exercise 7.2 :

Solve the following :-

(1) $(x^2 + 2x + 3) (2x^2 + 3x + 1)$

(2) $(2x^2 + 3x + 5) (x^2 + 2x + 2)$

(3) $(5x^2 - x + 3) (3x^2 + 5x - 2)$

(4) $(6x^2 + 3x - 5) (3x^2 + 2x - 5)$

(5) $(3x^2 + 8x + 9) (5x^2 - 7x - 1)$

(6) $(4x^2 - 2x - 6) (5x^2 + 3x + 2)$

(7) $(5x^2 + 3x - 2) (4x^2 - 2x - 1)$

(8) $(x^2 - x - 1) (2x^2 - x - 2)$

(9) $(x^2 - x - 1) (x^2 - x - 1)$

(10) $(x^2 + x + 1) (x^2 + x + 1)$

Special Cases

When we have to multiply two equations with unequal number of terms then we follow the same method of multiplication but with a little careful examination of the equations. We place zero as the coefficients for the variables, which are missing in the equation.

Example : $(2x^2 + x + 3)(x + 2)$

We note that the x^2 term is missing in the second equation, so when we are putting down our coefficients below our variables we place a zero below x^2 for the second equation.

Example 1 : $(2x^2 + x + 3)(x + 2)$

Step 1:

x^2	x	-
2	1	3
0	1	2

We place our coefficient below the variable taking care to place the zero below x^2.

Step 2:

x^2	x	-
2	1	3
0	1	2

$0 + 2 + 5 + 5 + 6$

Vertical and crosswise multiplication from the right.
(1) $3 \times 2 = +6$
(2) $(1 \times 2) + (3 \times 1) = +5$
(3) $(2 \times 2) + (0 \times 3) + (1 \times 1) = +5$
(4) $(2 \times 1) + (0 \times 1) = +2$
(5) $0 \times 2 = 0$

Step 3:

x^2	x	-
2	1	3
0	1	2

$2x^3 + 5x^2 + 5x + 6$

We add the variables to our coefficients in the answer. Since we have 0 as the coefficient for x^4 term, we don't write it as part of the answer.

Answer : $(2x^2 + x + 3)(x + 2)$
$= 2x^3 + 5x^2 + 5x + 6$

Vedic Mathematics for all Ages

Example 2 : $(x^2 + 3) (2x^2 + x + 2)$

In this example we do not have an 'x' term in the first equation.

Step 1:

x^2	x	-
1	0	3
2	1	2

We place our coefficients below our variables. We don't have x term in the 1st equation so we place a zero below x.

Step 2:

x^2	x	-
1	0	3
2	1	2

$2 + 1 + 8 + 3 + 6$

Vertical and crosswise multiplication from the right.
(1) $3 \times 2 = +6$
(2) $(0 \times 2) + (3 \times 1) = +3$
(3) $(1 \times 2) + (2 \times 3) + (0 \times 1) = +8$
(4) $(1 \times 1) + (0 \times 2) = +1$
(5) $2 \times 1 = 2$

Step 3:

x^2	x	-
1	0	3
2	1	2

$2x^4 + x^3 + 8x^2 + 3x + 6$

We now add the variables to the coefficients of the answer starting from the right.
We observe that even though we did not have one term in the first equation, we still have all the terms in the answer.
The placement of zero is in such a position that its product with other digits is always added to other products to give the coefficient, so it does not affect the variables in the answer.

Answer : $(x^2 + 3)(2x^2 + x + 2)$
$$= 2x^4 + x^3 + 8x^2 + 3x + 6$$

Example 3 : $(x^2 + x) (2x^2 + x + 3)$

In this example we do not have an independent term in the first equation.

Step 1:

x^2	x	-
1	1	0
2	1	3

We place our coefficients below our variables. As we don't have an independent term in the 1st equation we place a zero for it.

Step 2:

$$
\begin{array}{ccc}
x^2 & x & - \\
1 & 1 & 0 \\
2 & 1 & 3 \\
\hline
\end{array}
$$

$2 + 3 + 4 + 3 + 0$

Vertical and crosswise multiplication from the right.
(1) $3 \times 0 = 0$
(2) $(3 \times 1) + (0 \times 1) = +3$
(3) $(3 \times 1) + (2 \times 0) + (1 \times 1) = +4$
(4) $(1 \times 1) + (1 \times 2) = +3$
(5) $2 \times 1 = 2$

Step 3:

$$
\begin{array}{ccc}
x^2 & x & - \\
1 & 1 & 0 \\
2 & 1 & 3 \\
\hline
\end{array}
$$

$2x^4 + 3x^3 + 4x^2 + 3x + 0$

We add the variables to the coefficients of the answer. As we have 0 as the independent term we will not have an independent term in the answer.

Answer : $(x^2 + x)(2x^2 + x + 3)$
$= 2x^4 + 3x^3 + 4x^2 + 3x$

Example 4 : $(x^2 + 3) (2x^2 + x)$

In this example we do not have an x term in the first equation and an independent term in the second equation, so we will have to place two zeros when we place the variables.

Step 1:

$$
\begin{array}{ccc}
x^2 & x & - \\
1 & 0 & 3 \\
2 & 1 & 0 \\
\hline
\end{array}
$$

We place our coefficients below our variables. Since we don't have an x term in the 1st equation and independent terms in the 2nd equation, we place zeros for both.

Step 2:

$$
\begin{array}{ccc}
x^2 & x & - \\
1 & 0 & 3 \\
2 & 1 & 0 \\
\hline
\end{array}
$$

$2 + 1 + 6 + 3 + 0$

Vertical and crosswise multiplication from the right.
(1) $3 \times 0 = 0$
(2) $(3 \times 1) + 0 = +3$
(3) $(1 \times 0) + (2 \times 3) + (0 \times 1) = +6$
(4) $(1 \times 1) + (0 \times 2) = +1$
(5) $2 \times 1 = 2$

Step 3:

$$
\begin{array}{ccc}
x^2 & x & - \\
1 & 0 & 3 \\
2 & 1 & 0 \\
\hline
\end{array}
$$

$2x^4 + x^3 + 6x^2 + 3x$

We add the variables to the coefficients of the answer. We have zero as the independent term hence we don't have an independent term, and the other zero does not affect any other coefficient in the answer.

Answer : $(x^2 + 3)(2x^2 + x) = 2x^4 + x^3 + 6x^2 + 3x$

Example 5 : $(x^2 + x)(2x + 3)$

In this example we do not have an independent term in the first equation and an x^2 term in the second equation, so we will have to place two zeros when we place the variables.

x^2	x	-
1	1	0
0	2	3

$0 + 2 + 5 + 3 + 0$

We place our variables and their coefficients below them keeping in mind the two zeros. We then perform vertical and crosswise multiplication and then add the variables to get our answer.

Answer : $(x^2 + x)(2x + 3) = 2x^3 + 5x^2 + 3x$

Example 6 : $(4x^2 - 3x - 2)(2x - 4)$

x^2	x	-
4	-3	-2
0	2	-4

$0 + 8 - 22 + 8 + 8$

We place our variables and their coefficients below them and then perform vertical and crosswise multiplication and add the variables to get our answer.

Answer : $(4x^2 + 3x + 2)(2x + 4)$
$= 8x^3 - 22x^2 + 8x + 8$

Example 7 : $(3x^2 - 2x - 5)(2x^2 - 7)$

x^2	x	-
3	-2	-5
2	0	-7

$6 - 4 - 31 - 14 + 35$

We place our variables and their coefficients below them and perform vertical and crosswise multiplication and then add the variables to get our answer.

Answer : $(3x^2 + 2x - 5)(2x^2 - 7)$
$= 6x^4 - 4x^3 - 31x^2 - 14x + 35$

Example 8 : $(5x^2 - 3) (2x^2 - 5x + 2)$

x^2	x	-
5	0	-3
2	-5	2

$10 - 25 + 4 + 15 - 6$

We place our coefficients and their coefficients below them and perform vertical and crosswise multiplication and then add the variables to get our answer.

Answer : $(5x^2 - 3)(2x^2 - 5x + 2)$
$$= 10x^4 - 25x^3 + 4x^2 + 15x - 6$$

Rules :

(1) In any case of multiplication of two equations we first have to observe the variables.

(2) We write our variables and below them we write our coefficients of those terms for both the equations.

(3) We perform a simple vertical and crosswise multiplication.

(4) In the end we add the variables to the coefficients of our answer.

(5) We observe the numbers we multiplied to get the coefficients in the answer and then observe the variables of those coefficients. We multiply those variables and get the variable of that coefficient in the answer.

Exercise 7 .3 :

Solve the following equations :-

(1) $(x^2 + 2x + 1) (2x^2 + x)$

(2) $(x^2 - x + 1) (x + 2)$

(3) $(4x^2 + 2x + 3) (x - 5)$

(4) $(x^2 + 4x - 5) (x^2 + 3)$

(5) $(x^2 + 2) (x^2 + 5)$

(6) $(x + 2) (x^2 + 3)$

(7) $(x^2 + 3x - 7) (x^2 - x)$

(8) $(x^2 - 7) (x + 5)$

(9) $(2x^2 + 5x) (4x^2 + 2)$

(10) $(x^2 - 7x) (x + 5)$

Equations with Different Variables

Till now we have seen all different types of sums but they all had the same type of variables in each equation to be multiplied. But if both the equations to be multiplied have totally different variables, we still follow the same pattern but while adding the variable we have to take care in each term. Let us take an example to understand it.

Example 1 : $(4x + 7)(2z + 3y)$

Step 1:

x / z	- / y
4 (x)	7
2 (z)	3 (y)

We have two variables for each column. We place our coefficients below the variables. When we have different variables we keep writing the variables also along with our multiplication.

Step 2:

x / z	- / y
4 (x)	7
2 (z)	3 (y)

$8xz + 12xy + 14z + 21y$

Vertical and crosswise multiplication from the right.

(1) $(7 \times 3)y = 21y$ (we have multiplied independent term with a y term)

(2) $(4 \times 3) xy + (7 \times 2)z = 12xy + 14z$ (we multiplied x term with y term and independent term with a z term)

(3) $(4 \times 2) xz = 8xz$
(we multiplied x term with a z term)

Answer : $(4x + 7)(2z + 3y)$
$= 8xz + 12xy + 14z + 21y$

Here we see that we have performed our multiplication the same way as the previous examples but we have taken into account the variable when we perform our multiplication in each step. During our cross multiplication we will get two terms instead of the usual one term as we have different variables in both the columns.

Example 2 : $(x^2 + 2x + 6)(2y^2 + 3y - 2)$

Step 1:

x^2/y^2	x /y	-/-
1	+2	+6
2	+3	-2

We have two variables for each term. We place our coefficient below the variable. We keep writing the variables also along with our multiplication.

Step 2:

x^2/y^2	x/y	$-/-$
1	+2	+6
2	+3	-2

$2x^2y^2 + 3x^2y + 4y^2x + 6xy$
$-2x^2 + 12y^2 - 4x + 18y - 12$

Vertical and crosswise multiplication from the right.

(1) $(6 x -2) = -12$
   We have multiplied both independent terms

(2) $(2 x -2) x + (6 x 3) y = -4x + 18y$
   We multiplied an x term with an independent term and then a y term with an independent term.

(3) $(1 x -2) x^2 + (2 x 6) y^2 + (2 x 3) xy$
   $= -2 x^2 + 12 y^2 + 6xy$
   We multiplied an x^2 term with independent term, y^2 term with an independent term and then an x term to a y term.

(4) $(1 x 3) x^2 y + (2 x 2) y^2 x = 3 x^2 y + 4 y^2 x$
   We multiplied an x^2 term to a y term and y^2 term to an x term.

(5) $(2 x 1) x^2 y^2 = 2 x^2 y^2$
   We multiplied an x^2 term to a y^2 term.

Answer : $(x^2 + 2x + 6) (2y^2 + 3y - 2)$
   $= 2x^2y^2 + 3x^2y - 2x^2 + 12 y^2 + 6xy$
   $+ 4y^2x - 4x + 18y - 12$

Here we have different variables in all the columns except the last (independent term). So except for the first and the last term all the middle terms will have two or three terms with different variables in the answer. So instead of getting five terms we have nine terms in the answer.

Given below are a few more examples solving sums in straight steps :

Example 3 : $(4x+3z)(2a +6b)$

x/a	z/b
4	3
2	6

$8ax + 6az + 24 bx + 18 bz$

Answer : $8ax + 6az + 24 bx + 18 bz$

Example 4 : $(2a^2 - 3b + c)(4x^2+3x -2)$

a^2/x^2	b/x	$c/-$
2	-3	1
4	3	-2

$8a^2x^2 + 6a^2x - 12bx^2 - 4a^2 + 4x^2c - 9bx$
$+ 3x + 6b - 2c$

Answer : $8a^2x^2 + 6a^2x - 12bx^2 - 4a^2 + 4x^2c - 9bx + 3x + 6b - 2c$

Exercise 7.4 :

Multiply the following equations having different variables :-

(1) $(2x + 4)(4y + 5)$

(2) $(3x + 4)(2y - 7)$

(3) $(2x - 3)(5x - 7y)$

(4) $(6x - 2)(4 + 5y)$

(5) $(2x^2 + x + 3)(y^2 + 3y + 3)$

(6) $(x^2 + x + 2)(2y^2 + y - 4)$

(7) $(3x^2 - x + 1)(y^2 + y - 1)$

(8) $(2x^2 - 3y + 5)(4y^2 + 6x + 8)$

Answers

Exercise 7.1

(1) $6a^2 + 17ab + 12b^2$ (2) $21a^2 + 29ab - 10b^2$ (3) $8x^2 + 14xy + 3y^2$

(4) $35x^2 + 18xy - 8y^2$ (5) $34x^2 - 113xy - 21y^2$ (6) $138x^2 + 16xy - 10y^2$

(7) $33a^2 - 105ab + 18b^2$ (8) $15x^2 - 39xy - 18y^2$ (9) $21p^2 - 41pq + 18q^2$

(10) $85m^2 - 99mn - 28n^2$

Exercise 7. 2

(1) $2x^4 + 7x^3 + 13x^2 + 11x + 3$ (2) $2x^4 + 7x^3 + 15x^2 + 16x + 10$

(3) $15x^4 + 22x^3 - 6x^2 + 17x - 6$ (4) $18x^4 + 21x^3 - 39x^2 - 25x + 25$

(5) $15x^4 + 19x^3 - 14x^2 - 71x - 9$ (6) $20x^4 + 2x^3 - 28x^2 - 22x - 12$

(7) $20x^4 + 2x^3 - 19x^2 + 1x - 2$ (8) $2x^4 - 3x^3 - 3x^2 + 3x + 2$

(9) $x^4 - 2x^3 - x^2 + 2x + 1$ (10) $x^4 + 2x^3 + 3x^2 + 2x + 1$

Exercise 7. 3

(1) $2x^4 + 5x^3 + 4x^2 + x$ (2) $x^3 + x^2 - x + 2$ (3) $4x^3 - 18x^2 - 7x - 15$

(4) $x^4 + 4x^3 - 2x^2 + 12x - 15$ (5) $x^4 + 7x^2 + 10$ (6) $x^3 + 2x^2 + 3x + 6$

(7) $x^4 + 2x^3 - 10x^2 + 7x$ (8) $x^3 + 5x^2 - 7x - 35$ (9) $8x^4 + 20x^3 + 4x^2 + 10x$

(10) $x^3 - 2x^2 - 35x$

Exercise 7. 4

(1) $10x + 16y + 8xy + 20$ (2) $8y - 21x + 6xy - 28$

(3) $10x^2 - 15x + 21y - 14xy$ (4) $24x - 10y + 30xy + 8$

(5) $2x^2y^2 + xy^2 + 6x^2y + 3xy + 6x^2 + 3y^2 + 3x + 9y + 9$

(6) $2x^2y^2 + 2xy^2 + x^2y + xy - 4x^2 + 4y^2 - 4x + 2y - 8$

(7) $-3x^2y^2 - xy^2 + 3x^2y - xy - 3x^2 + y^2 + x + y - 1$

(8) $8x^2y^2 - 12y^3 + 12x^3 - 18xy + 20x^2 + 16y^2 + 30x - 24y + 40$

Chapter 8

DIGITAL ROOTS

Vedic methods give us answers in straight steps and they are very accurate, easy and fast, but we also have Digital Root which helps us to check the accuracy of our answers in a fraction of a second without redoing the whole sum again. Digital roots or **"Navshesh"** of any number is the sum of its digits continued till we get only a single digit.

Let us take a few examples to understand this.

Example : Digital Roots of the following are :

412 = 4 + 1 + 2 = 7
654 = 6 + 5 + 4 = 15 = 1 + 5 = 6
3861 = 3 + 8 + 6 + 1 = 18 = 1 + 8 = 9

There is a simpler method of finding out the digital root of bigger numbers. Instead of adding all the digits in the number we find out the digital roots by *casting out nines*. We cancel out all nines in the given numbers, and the number which add up to nine. We then add up the remaining digits and convert them into a single digit, and this gives us the digital root of the number.

Example :

4819456 = We can cancel out 9, (8, 1), (4, 5)
           We are left with 4, 6
           Digital root = 4 + 6 = 10 = 1+ 0 = 1

7425241 = We can cancel out (7, 2), (4, 5)
           We are left with 2, 4, and 1
           So digital root is 2 + 4 + 1= 7

81945 = We can cancel out (8, 1), 9, (4, 5)
         After cancelling out all digits we don't have any digits left with us, In this case the digital root is not zero but 9. Since digital root cannot be zero for any number.

Exercise 8 .1 :

Find the digital root of the following :-

(1) 64729   (6) 24851

(2) 64591824   (7) 3578213

(3) 86221   (8) 72365485

(4) 32812   (9) 21583

(5) 996521   (10) 984135

Application of Digital Roots

To verify our answers for any problem we use digital roots, we perform the same mathematical operation on digital roots of the numbers as done in the problem, and then the answer we get should be equal to the digital roots of the answer of the sum. If the answer we get is not equal to the digital root of the answer of the original problem then our answer is not correct. Let us understand how digital roots help to verify our answers in sums with different mathematical operations with the help of a few examples.

In the case of Addition :

If the sum of the digital roots of the numbers is equal to the digital root of the answer then the answer is correct.

$$\begin{array}{r} 5\,6\,7 \\ +\ 2\,7\,4 \\ \hline 8\,4\,1 \\ \hline \end{array}$$

Digital root of 567 = 9
Digital root of 274 = 4

We ADD both the digital roots = 9 + 4 = 13
Digital root of 13 = 4

And digital root of 841 = 4

This verifies the answer as correct.

In the case of Subtraction :

If the difference of the digital roots of the numbers is equal to the digital roots of the answer then our answer is correct.

$$
\begin{array}{r}
462 \\
-134 \\
\hline
328 \\
\hline
\end{array}
$$

Digital root of 462 = 3
Digital root of 134 = 8

We SUBTRACT both the digital roots
= 3 - 8 = -5 = 4
(Digital root of a number cannot be negative so we take the complement of the number from 9)

And digital root of 328 = 4

This verifies the answer as correct.

In the case of Multiplication :

If the product of the digital roots of the two numbers is equal to the digital root of the answer then our answer is correct.

$$
\begin{array}{r}
273 \\
\times \ 384 \\
\hline
104832 \\
\hline
\end{array}
$$

Digital root of 273 = 3
Digital root of 384 = 6

We MULTIPLY both the digital roots
= 3 × 6 = 18 = 9

And digital root of 104832 = 9

This verifies the answer as correct.

In the case of Division :

If the product of the digital root of the Divisor and Quotient when added to the digital root of the remainder is equal to the digital root of the Dividend then our answer is correct.

Digital root of (quotient) 7211 = 2
Digital root of (divisor) 8 = 8
$5\,7\,6\,9\,3\,/\,8$ Digital root of (remainder) 5 = 5

gives   We multiply the digital roots of quotient and
Quotient = 7211   divisor and add digital root of remainder
Remainder = 5 $= 2 \times 8 + 5 = 21 = 3$
                         And Digital root of (dividend) 57693 = 3

This verifies the answer as correct.

<u>IMPORTANT :</u> This digital root method does not hold true if the digits in the answer are interchanged. As , when this happens the digital root will remain the same but the answer will be wrong.

Exercise 8 .2 :

Check the answers to the following sums using digital roots :-

(1) 8563 + 6575   (6) 68 × 93

(2) 85647 + 74429   (7) 956 × 164

(3) 9354 – 2754   (8) 985/7

(4) 64843 – 16763   (9) 8337/21

(5) 73372 - 4773   (10) 35465/32

Answers

Exercise 8

(1) 1	(2) 3	(3) 1	(4) 7	(5) 5	(6) 2	(7) 2
(8) 4	(9) 1	(10) 3				

Chapter 9

DIVISIBILITY

All numbers are divisible by some number or the other. 1 is not divisible by any number so it is indivisible. The numbers which are only divisible by themselves and 1 (one) are called Prime numbers. All other numbers are called composite numbers.

Now let us try and learn the divisibility of some numbers.

Divisibility by 2 :

All even numbers, which end in 2, 4, 6, 8, 0 are divisible by 2.
Example : 3756, 275648, 93766854, all these numbers are even and also divisible by 2.

Divisibility by 5 :

All numbers that end in 5 or 0 are divisible by 5.
Example : 45, 92630, 756595, 8394725 are completely divisible by 5 as all end in either 5 or zero.

Divisibility by 10 :

All numbers that end in 0 are divisible by 10.
Example : 193750, 343500, 456000 are completely divisible by 10.

Divisibility by 3 :

To find if a number is divisible by 3, we add all the digits and if the sum is divisible by 3 then the number is divisible by 3.

Example : Is 67495 divisible by 3?

> The digits add up to **31** (6+7+4+9+5 = 31),
> 31 is not divisible by 3
> So 67495 is NOT divisible by 3.

<u>Example</u> : Is 63972 divisible by 3?

The digits add up to 27 (6+3+9+7+2 = 27),
27 is divisible by 3, So 63927 is completely divisible by 3.

Divisibility by 9 :

Any number is divisible by 9 if the sum of its digits is divisible by 9.
We add up all the digits of the number and if the sum is divisible by 9
then the number is divisible by 9.

There is a short or and an easier way, by casting out 9's (as learnt in the
previous chapter) or cancel all 9's and digits adding up to 9, in the
number. Then we add up the remaining digits, and keep doing that till
we get a single digit number, if we get a zero or a nine then the number is
divisible by 9.

<u>Example</u> : Is 9456512 divisible by 9?

We cancel out all 9's and all number adding up to 9 i.e. 9, (4, 5),
(6, 1, 2)
We now have 5, so the number is NOT completely divisible by 9

<u>Example</u> : Is 36792225 divisible by 9?

We cancel out all 9's
We cancel out (3, 6), (7, 2), (5, 2, 2), 9
We now have 0 left, so the number is completely divisible by 9.

Divisibility by 4 :

If the last two digits of the number are divisible by 4 then the number is
divisible by 4.
Or if the sum of the last digit and twice the second last digit is divisible
by 4 then the number is divisible by 4.

<u>Example</u> : Is 476543 divisible by 4?

The sum of the last and twice the second last is
3 (last) + 2 × 4 (second last digit) = 11,
11 is not divisible by 4,
so 476543 is NOT divisible by 4.

Example : Is 39756 divisible by 4?

> The sum of the last and twice the second last is
> 6 (last) + 2 × 5 (second last digit) = 16,
> 16 is completely divisible by 4,
> so 476543 is completely divisible by 4.

Divisibility by 8 :

If the last three digits of the number are divisible by 8 then the number is divisible by 8.
Or if the sum of the last plus twice the second last and four times the third last digit is divisible by 8 then the number is divisible by 8.

Example : Is 83546 divisible by 8?

> Last + twice second last + four times third last is
> 6 + (2 × 4) + (4 × 5) = 34
> 34 is not divisible by 8,
> so 83546 is NOT completely divisible by 8.

Example : Is 235472 divisible by 8 ?

> Last + twice second last + four times third last is
> 2 + (2 × 7) + (4 ×4) = 32
> 32 is divisible by 8,
> so 235472 is completely divisible by 8.

Divisibility by 11 :

If the difference of the sum of the evenly placed digit from the sum of the oddly placed digits or vice versa gives zero, 11 or a multiple of 11 then the number is divisible by 11.

Example : Is 7834531 divisible by 11?

> Evenly placed digits are 8, 4, 3; their sum is 8 + 4 + 3 = 15
> Oddly placed digits are 7, 3, 5, 1; their sum is 7 + 3 + 5 + 1 = 16
> 16 – 15 = 1
> So 7834531 is NOT divisible by 11.

<u>Example</u> : Is 3544783 divisible by 11?

> Evenly placed digits are 8, 4, 5; their sum is 8 + 4 + 5 = 17
> Oddly placed digits are 3, 7, 4, 3; their sum is   3 + 7 + 4 + 3 = 17
> 17 – 17 = 0
> So 3544783 is completely divisible by 11.

Divisibility by Prime Numbers :

To check if the given number is divisible by any prime number we use its *ekadhika*. We multiply the last digit of the number by its *ekadhika* and add it to the other digits prior to the last digit. We keep doing it till we reach to a number that we already know is divisible by the given number.

Ekadhika of any prime number can be calculated by multiplying it by such a number that gives 9 in the units place, and then we take one more than the previous digit(s) to 9 of the answer, as the *ekadhika* of the number.

Ekadhika of 7 = 7 x 7 = 49, 4 (previous digit) +1 = 5
Ekadhika of 13 = 13 x 3 = 39, 3 (previous digit) +1 = 4
Ekadhika of 19 = 19 x 1 = 19, 1 (previous digit) +1 = 2

Divisibility by 7 :

Now we use the *ekadhika* to find the divisibility by 7
The *ekadhika* of 7 is 5.
We multiply the last digit of the number by its *ekadhika* and add it to the other digits prior to the last digit. We keep doing it till we reach a number that we already know is divisible by 7.

<u>Example</u> : Is 743693 divisible by 7 ?

> *Ekadhika* of 7 is 5
> Last digit is 3 so 3 x 5 = 15, we add this to 74369 = 74384
> Last digit now is 4 so 4 x 5 = 20, 7438 + 20 = 7458
> Last digit now is 8 so 8 x 5 = 40, 745 + 40 = 785
> Last digit now is 5 so 5 x 5 = 25, 78 +25 = 103
> Last digit now is 3 so 3 x 5 = 15, 10 +15 = 25
> 25 is not divisible by 7 so 743693 is not divisible by 7

Divisibility by 13 :

Example : Is 126594 divisible by 13 ?

Ekadhika of 13 = 13 x 3 = 39, 3(previous digit) +1 = 4

Last digit is 4 so 4 x 4 = 16, we add this to 12659 = 12675
Last digit now is 5 so 5 x 4 = 20, 1267 + 20 = 1287
Last digit now is 7 so 7 x 4 = 28, 128 + 28 = 156
Last digit now is 6 so 6 x 4 = 24, 15 + 24 = 39

39 is divisible by 13 so 126594 is divisible by 13

Divisibility rules for composite numbers :

To obtain the rules for divisibility of composite numbers or non prime numbers, we combine the rules of divisibility of their factors. For example 6 has its factors as 2 and 3 so for a number to be divisible by 6 it should be divisible by 2 and 3.

Some of the rules are as follows :-

(1) A number is divisible by 6 when it is divisible by both 2 and 3.
(2) A number is divisible by 12 when it is divisible by both 3 and 4.
(3) A number is divisible by 15 when it is divisible by both 3 and 5.
(4) A number is divisible by 18 when it is divisible by both 2 and 9.
(5) A number is divisible by 24 when it is divisible by both 3 and 8.
(6) A number is divisible by 30 when it is divisible by both 3 and 10.
(7) A number is divisible by 20 when it ends in zero and second last digit is even.

Example : Is 1379112 divisible by 24 ?

If the number is divisible by 3 and 8 then the number is divisible by 24

To check the divisibility by 3 we add all the digits
1+3+7+9+1+1+2 = 24 = 2+4 = 6
6 is divisible by 3 so 1379112 is divisible by 3

To check the divisibility by 8 we add the last plus twice second last plus four times third last digits

$2 + 2 \times 1 + 4 \times 1 = 8$
So 1379112 is divisible by 8 also

Since 1379112 is divisible by both 3 and 8 it is divisible by 24.

Exercise 9 :

Which of the following are divisible by the given number ?

(1) 4321, 7865, 6745, 63963 by 3

(2) 536, 8765, 48654, 4864 by 5

(3) 540, 5642,78546,64678 by 6

(4) 5436,79492,78654,65846 by 7

(5) 3246,5368,2178,87321 by 8

(6) 76539, 95346,476325,7643254 by 9

(7) 675334,56487,39765,648768 by 11

(8) 9870,12345,6545,987080 by 15

(9) 876,187632,756432,76546 by 18

(10) 4624,875432,841128,78698 by 24

Answers

Exercise 9

(1) No, No, No, Yes   (2) Yes, No, No, Yes   (3) Yes, No, Yes, No
(4) No, Yes, No, No   (5) No, Yes, No, No   (6) No, Yes, Yes, No
(7) Yes, No, Yes, No   (8) Yes, Yes, No, No   (9) No, Yes, Yes, No
(10) No, No, Yes, No

Sunflower Bloom

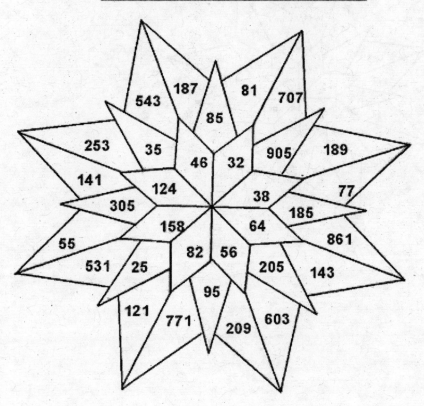

If the number is divisible by :	Colour the shape :
2	Yellow
3	Red
5	Orange
11	Dark Red

Answers to Sunflower Bloom

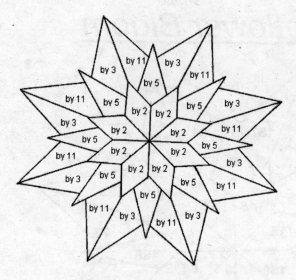

Chapter 10

DIVISION - I

As we have seen in the previous chapters that in Vedic mathematics there are different methods of calculations for different types of numbers. We have learnt various ways of multiplication like those for numbers close to the base and those for random numbers. In division also there are many different methods for those numbers close to the bases and those for random numbers. We will first learn division by numbers close to the bases. In this chapter we will deal with numbers less than the base. For this we use the sutra

निखिलं नवतश्चरमं दशतः

"Nikhilaṃ Navataścaramaṃ Daśataḥ"

"All from nine and last from ten"

At the beginning let us start division by the number 9. We can then apply the same method for the division by big numbers close to the bases. Division by 9 is very easy and interesting.

Let us understand it with the help of a few examples

<u>**Example 1**</u> : 103/9

<u>Step 1:</u> 9 | 1 0 / 3	We write the number as shown Our divisor is 9 and its base is 10 so we will have 1 digit in the remainder as (10 has one zero). To separate our remainder we put a slash before 3.
<u>Step 2:</u> 9 | 1 0 / 3 1	We bring down the first digit of the given number as the first digit of the quotient. We place our quotient below the line.

Step 3:

We now multiply this quotient digit with the complement of our divisor,

complement of 9 = 1

1(quotient digit) x 1(complement) = 1

so we get the same digit and we then place this digit under the second digit of our number (0)

Step 4:

9 | 1 0 / 3
     1   1

1 1

We add up the digits in the second column and write the sum as the second digit of the quotient

0 + 1 = 1

We now multiply the new quotient digit with the complement of the divisor

1 x 1 (complement of 9) = 1

place it below the third digit of the dividend

Step 5:

9 | 1 0 / 3
       1   1

1 1 / 4

Now since we are in the remainder side, we add the digits in the remainder. The addition in the remainder side will give us the remainder of the division.

3 + 1 = 4

4 is the remainder of our division.

Answer : 103/9 gives Quotient (Q) = **11** and
Remainder (R) = **4**

Example 2 : 124/9

Step 1:

9 | 1 2 / 4

We write the number as shown

Our divisor is 9 and its base is 10, so we will have 1 digit in the remainder. To separate our remainder we put a slash before 4.

Step 2:

9 | 1 2 / 4
      1

We bring down the first digit of the number as the first digit of the quotient.

We place our quotient below the line

Step 3:

```
9 | 1 2 / 4
  |   1
  |   ↗
    1
```

We now multiply this quotient digit with the complement of our divisor,
   complement of 9 = 1
   1 x 1 = 1
so we get the same digit and we then place this digit under the second digit of our number (2).

Step 4:

```
9 | 1 2 |/ 4
  |   1 ↓ 3
  |     ↗
    1 3
```

We add up the digits in the second column and write the sum as the second digit of the quotient.
   2 + 1 = 3

We now multiply the new quotient digit with the complement of the divisor

   3 x 1 (complement of 9) = 3
Place it below the third digit of the dividend(4).

Step 5:

```
9 | 1 2 / 4 |
  |   1   3 ↓
  |
    1 3 / 7
```

Now since we are in the remainder side, we now add the digits in the remainder. The addition in the remainder side will give us the remainder of the division.

   4 + 3 = 7
This is the remainder of our division.

Answer : 124/9 gives **Q = 13 , R= 7**

Example 3 : 2211/9

Step 1:

```
9 | 2 2 1 / 1
```

We write the number as shown.
Our divisor is 9 and its base is 10, so we will have 1 digit in the remainder. To separate our remainder we put a slash before 1.

Step 2:

```
9 | 2 | 2 1 / 1
  |   ↓
    2
```

We bring down the first digit of the number as the first digit of the quotient.

We place our quotient below the line

Step 3:

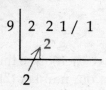

2

We now multiply this quotient digit with the complement of our divisor,
    complement of 9 = 1
    2 x 1 = 2
So we get the same digit and we then place this digit under the second digit of our number (2)

Step 4:

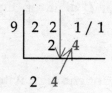

2   4

We add up the digits in the second column and write the sum as the second digit of the quotient.

    2 + 2 = 4
We now multiply the new quotient digit with the complement of the divisor.

    4 x 1 (complement of 9) = 4
Place it below the third digit of the dividend (1).

Step 5:

9 | 2 21 | / 1
    24   5

2 4 5 /

We now add the digits in the third column.
    1 + 4 = 5
This is the third digit of the quotient.
    5 x 1 (complement of 9) = 5

and place this below the remainder digit (1).

Step 6:

9 | 2 2 1 / 1
     2 4   5

2 4 5 / 6

We now add the digits in the remainder.

    1 + 5 = 6

This is the remainder of our division.

Answer : 2211/9 gives **Q = 245** , **R = 6**

Example 4 : 237/9

Step 1:

9 | 23 / 7

We write the number as shown.
Our divisor is 9 and its base is 10, so we will have 1 digit in the remainder. To separate our remainder we put a slash before 4.

Step 2:

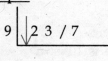

We bring down the first digit of the number as the first digit of the quotient.
We place our quotient below the line

Step 3:

9 | 2 3 / 7
     2
  2

2×1 (complement of 9) = 2
We now place this digit under the second digit of our number (3)

Step 4:

9 | 2 3 | / 7
     2   5
  2   5

We add up the digits in the second column and write the sum as the second digit of the quotient.
    3 + 2 = 5
    5×1 (complement of 9) = 5
We now place it below the third digit of our number (7).

Step 5:

9 | 2 3 / 7
     2   5
  2   5 / 12

We now add the digits in the remainder.
    7 + 5 = 12
Here we see that the remainder is bigger than the divisor.
So we again divide the remainder by our divisor 9.

Step 6:

9 | 2 3 / 7
     2   5
  2   5 /

    1 / 2
       1
    1 / 3

We again keep our remainder digit (2) aside.
We bring down the first digit (1) of the new dividend (12) as the new quotient.

Now 1×1 (complement of 9) = 1
We write it under our next digit (2)
Now we add the two digits
    2 + 1 = 3

This is the new remainder of our division
We add the quotient (1) of this part to our original quotient (25)
Q = 25 + 1 = 26
R = 3
Answer : 237/9 gives Q = 26 , R = 3

Example 5 : 21451/9

Step 1:

9 | 2 1 4 5 / 1

We write the number as shown.
Our divisor is 9 and its base is 10, so we will have 1 digit in the remainder. To separate our remainder we put a slash before 1.

Step 2:

9 | 2 |1 4 5 / 1

2

We bring down the first digit of the number as the first digit of the quotient.
We place our quotient below the line.

Step 3:   2 x 1 (complement of 9) = 2

9 | 2   1 4 5 / 1
        2

2

We then place this digit under the second digit of our number (1).

Step 4:

9 |2 1 | 4 5 /1
      2   3

    2 3

We add up the digits in the second column and write the sum as the second digit of the quotient.
            1 + 2 = 3
            3 x 1 = 3
place it below the third digit of the dividend(4).

Step 5:

9 |2 1 4 | 5 / 1
      2 3   7

  2 3 7

We now add the digits in third column
            4 + 3 = 7
This is the third digit of the quotient
            7 x 1 = 7
place it below the fourth digit of the dividend(5).

Step 6:

9 | 2 1 4 5 |/ 1
      2 3 7

2 37 ₁2
    ⤳

We now add the digits in fourth column
            5 + 7 = 12
This is the third digit of the quotient.
But 12 is a double digit number so for the quotient part we place 2 and *carry over* 1 to the previous digit of quotient (left side) i.e. 7
            7 + 1 = 8
quotient becomes 2382

Step 7:

```
9 | 2 1 4 5 / 1 |
  |   2 3 7   12 ↓
  
    2 3 8 2  / 13
```

But we carry over the whole number 12

12 x 1 (complement) = 12

to be added to the next digit of the dividend
We now add the digits in the remainder

1 + 12 = 13

This is the remainder of our division but it is bigger than 9 so we divide it again.

We again keep our remainder digit (3) aside.

Step 8:

```
9 | 2 1 4 5 / 1
  |   2 3 7   12
  
    2 3 8 2 / 13
             |
         1 / 3 |
             1 ↓
        ‾‾‾‾‾‾‾
         1 / 4
```

We bring down the first digit (1) of the new dividend (13) as the new quotient

1 x 1 (complement) = 1

We write it under our next digit (3)
Now we add the two digits

3 + 1 = 4

This is the new remainder of our division.
We add the quotient (1) of this part to our original quotient (2382).

2382 + 1 = 2383

Answer : 21451/9 gives **Q = 2383 , R = 4**

Example 6 : 1011649/9

```
9 | 1 0 1 1 6 4 / 9
  |   1 1 2 3 9  13
  ‾‾‾‾‾‾‾‾‾‾‾‾‾‾‾‾‾‾
    1 1 2 3 9 13 / 22

  1 1 2 3 9 3  /     2 / 2
          1          2
  ‾‾‾‾‾‾‾‾‾‾‾‾    ‾‾‾‾‾‾‾‾‾
  1 1 2 4 0 3       2 / 4
```

(1) Bring down 1
(2) Place it under 0(second digit of dividend), and add the digits
   0 + 1 = 1 (this is the next digit of Q)
(3) Place it under 1 (third digit of dividend) and add
   1 + 1 = 2 (this is the next digit of Q)
(4) Place it under 1 (fourth digit of dividend) and add
   1 + 2 = 3 (next digit of Q)
(5) Place it under 6 (fifth digit of dividend) and add
   6 + 3 = 9 (next digit of Q)
(6) Place it under 4 (sixth digit of

dividend) and add
4 + 9 = 13 (next digit of Q)

But 13 is a double digit number so for the quotient part we place **3** and *carry over* 1 to the previous digit of the quotient (9)

(7) However, we place 13 itself under 9 (remainder part) and add
9 + 13 = 22
(remainder, but it is bigger than the divisor)

(8) We divide 22 again by 9
bring down 2 and carry it over, to be added to 2 (the next digit of the new dividend 22)

We get remainder 4
We get Q = 2 and R = 4

This new quotient (2) is added to the other quotient (11203) hence

Final Q = 112403 + 2 = **112405**,
R = **4**

Answer : 1011649/9 gives
Q = 112405 , R = 4

Rules :

(1) The first digit of the dividend becomes the first digit of the quotient.

(2) It is multiplied with the complement of the divisor. In the case where divisor is 9 and the complement is 1 the first digit of the

quotient is carried and added to the second digit of the dividend. This addition gives the second digit of the quotient.

(3) The second digit of the quotient is added to the third digit of the dividend, which gives the third digit of the quotient.

(4) We carry the same procedure till we reach the last digit or the remainder of the dividend.

(5) If we get a double digit in the addition of any of the columns, then

- We carry it as it is to the next column for further calculations,

- In the quotient part we put down the unit digit and carry over the other digits to the previous digit (left side) of the quotient.

(6) The addition in the remainder part is the remainder of the sum.

(7) If we get a number bigger than the divisor then, in the remainder side we further divide it

- Add the quotient of this part to the previous quotient,

- The remainder of this part becomes the remainder of the sum.

Exercise 10 .1 :

Attempt the following divisions :-

(1) 213/9

(2) 12301/9

(3) 1200212/9

(4) 136/9

(5) 235/9

(6) 225/9

(7) 5201/9

(8) 42014/9

(9) 35216/9

(10) 51342/9

Division by Numbers close (less) than the Base

Now let us extend our division to numbers greater than 9. We will first deal with numbers close to the bases. We have observed that when we divide any number by 9 we multiply the quotient digit by the complement of the divisor (in this case it was one) and add it to the next dividend digit.

Let us apply the same formula to division by big numbers close to the bases. In this type of division we convert our big numbers to smaller numbers by taking their complement. This complement becomes our new divisor. This whole method makes the division very simple and interesting. It even encourages us to look at any problem in a different way.

Here we use the sutra *"Nikhilam Navtascharmam Dastah"* to find out the complement of the divisor. We carry on the division with this new divisor in a different manner.

Let us take a few examples to understand the working:

Example 1 : 123/89

Step 1:

89 | 1 / 23
11 |_____

We write our number as shown.
Here we have 89 as the divisor, which has its base as 100 so we will have 2 digits in the remainder.
We separate the last two digits as the remainder part.
We will now find the complement of 89
    complement of 89 = 11

This is our new divisor.

Step 2:

89 | 1 / 2 3
11 |_____
       1

Here we bring down the first digit of the dividend as the first digit of the quotient as done in the previous divisions.

Step 3:

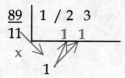

In the next step we multiply the first digit of the quotient by the new divisor
$$1 \times 11 = 11$$
and place it below the next digits of the dividend as shown
Only one digit in each column.
Units of 11, which is 1 is placed under 3 ,
Tens of 11, which is also 1 here, is placed under 2.

Step 4:

```
89 │ 1 / 2 │ 3 │
11 └      1 │ 1 ┘

     1 / 3   4
```

Since we are in the remainder side we will just add the digits in both the columns starting from the right.
$$3 + 1 = 4$$
$$2 + 1 = 3$$

So 34 becomes our remainder.
Here we have to see that our remainder has to be smaller than our original divisor but not the new divisor (complement of the divisor), 34 is smaller then 89 and hence our division is complete.

Answer : 123/89 gives **Q = 1 , R = 34**

Example 2 : 1234/888

Step 1:

```
888 │ 1 / 2 3 4
112 └
```

We write our number as shown. Here we have 888 as the divisor which has its base as 1000 so we will have 3 digits in the remainder
We separate 234 as remainder part.
We will now find the complement of 888.
complement of 888 = 112
This is our new divisor.

Step 2:

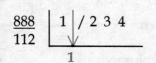

Here we bring down the first digit of the dividend as the first digit of the quotient as done in the previous divisions.

Step 3:

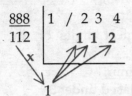

In the next step we multiply the first digit of the quotient by the new divisor and place it below the digits of the dividend as shown. Only one digit in each column.

$1 \times 112 = 112$

Units of 112, which is 2, is placed under 4, Tens of 112, which is 1, is placed under 3, Hundredth of 112, which is 1, is placed under 2.

Step 4:

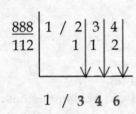

Since we are in the remainder side we will just add the digits in all the columns in the remainder side starting from the right.

$4 + 2 = 6$
$3 + 1 = 4$
$2 + 1 = 3$

So **346** becomes our remainder, as it is less then 888, our division is complete.

Answer : 1234/888 gives **Q = 1 , R = 346**

Example 3 : 12325/887

Step 1:

887 | 12/325
113 |

We write our number as shown. Here we have 887 as the divisor, which has its base as 1000 so we will have 3 digits in the remainder. 325 is in the remainder side.

complement of 887 = **113**
This is our new divisor.

Step 2:

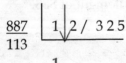

Here we bring down the first digit of the dividend as the first digit of the quotient as done in the previous divisions.

Step 3:

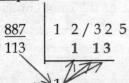

We multiply the first digit of the quotient by the new divisor and place it below the digits of the dividend as shown.

Only one digit in each column
$1 \times 113 = 113$

113 is placed under the digits of 232 of the dividend as shown, we have written digits only under 2 digits of the remainder side and not all three so we are still not in the remainder side.

Step 4:

$$\underline{887} \quad \begin{array}{c|c} 1\,2 & /\ 3\,2\,5 \\ 1 & \ \ 1\,3 \end{array}$$

$$1\,3$$

So we will just add the digits in one column only even though we have written digits in three columns.

$$2 + 1 = 3$$

This is the next digit of the quotient.

This way we keep adding digits of each column at a time and we get digits of the quotient in each step till we are in the remainder side.

Step 5:

$$\underline{887} \quad \begin{array}{c|c} 1\,2 & /\,3\ 2\ 5 \\ 1 & 1\ \ 3 \\ & 3\ 3\ 9 \end{array}$$

$$x$$

$$1\ 3$$

We now multiply the new digit of the quotient with the new divisor and place them below the next digits in the dividend starting from the third digit as shown.

$$3 \times 113 = 339$$

339 is placed under digits of 325 of the dividend as shown.

Step 6:

$$\underline{887} \quad \begin{array}{c|c|c} 1\,2/\,3 & 2 & 5 \\ 1\ \ 1 & 3 & \\ 3 & 3 & 9 \end{array}$$

$$1\,3\ /\ 7\ 9\ 4$$

Now since we are in the remainder side and we have written at least one digit below all our digits in the dividend, we will add all the digit column wise starting from the right.

$$5 + 9 = {}_{1}4$$
$$2 + 3 + 3 = 8 + 1\ (carry) = 9$$
$$3 + 1 + 3 = 7$$

Our remainder = 794

Answer : 12325/887 gives **Q = 13 , R = 794**

Example 4 : 2123/97

Step 1:

97 | 2 1 / 2 3
03 |

We place our number as shown. Here we have 97 as the divisor, which has its base as 100 so we will have 2 digits in the remainder.

complement of 97 = 03
Since 100 is the base we should have 2 digits in the complements
This is our new divisor.

Step 2:

97 | 2 | 1 / 2 3
03 |
    2

The first digit of the dividend is the first digit of the quotient.

Step 3:

97 | 2 1 / 2 3
03 | 0   6
 ×
        2

We multiply the first digit of the quotient by the new divisor and place it below the digits of the dividend as shown.
Only one digit in each column
      2 × 03 = 06
06 is placed under 12 of the dividend as shown.

Step 4:

97 | 2 1 | / 2 3
03 | 0   6
   |
       2 1

Since we are not in the remainder side we will just add the digits in one column only.

1 + 0 = 1
This is the next digit of the quotient

Step 5:

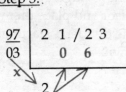

97 | 2 1 / 2 3
03 | 0   6
   | 0   3
 ×
      2   1

We now multiply the new digit of the quotient with the new divisor and place them below the next digits in the dividend starting from the third digit as shown.
      1 × 03 = 03

03 is placed under 23 of the dividend as shown.

Step 6:

```
97 | 21 / 2 | 3 |
03 |    0  6 |   |
   |       0  3 |
   |             |
   21 / 8  6
```

Now since we are in the remainder side we will add all the digits column wise starting from the right.

$3 + 3 = 6$

$2 + 6 + 0 = 8$

Our Remainder = 86

Answer : 2123/97 gives **Q = 21 , R = 86**

Example 5 : 111111/9979

Step 1:

```
9979 | 11 / 1111
0031 |_____
```

We place our number as shown. Here we have 9979 as the divisor, which has its base as 10000 so we will have 4 digits in the remainder.

complement of 9979 = 0031

Since 10000 is the base we should have 4 digits in the complements

This is our new divisor.

Step 2:

```
9979 | 1 | 1 / 1 1 1 1
0031 |___↓_____
       1
```

First digit of the dividend is the first digit of the quotient.

Step 3:

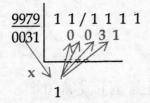

The first digit of the quotient is multiplied by the new divisor and placed below the digits of the dividend as shown.

Only one digit in each column

$1 \times 0031 = 0031$

If we take the complement as 31 instead of 0031 then our addition of all the columns will be different and we will not get the correct answer. So taking the correct complement is most important.

Step 4:

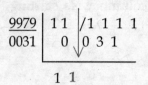

We will add the digits of one column only to get the next digit of the quotient.

$$1 + 0 = 1$$

This is the next digit of the quotient.

Step 5:

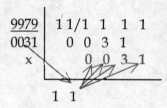

Again we multiply the new digit of the quotient with the divisor and place them below the next digits in the dividend starting from the third digit as shown.

$$1 \times 0031 = 0031$$

Step 6:

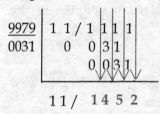

Now since we are in the remainder side we will add all the digits column wise starting from the right.

$$1 + 1 = 2$$
$$1 + 1 + 3 = 5$$
$$1 + 3 + 0 = 4$$
$$1 + 0 + 0 = 1$$

Our Remainder = 1452

Answer : 111111/9979 gives
Q = 11 , R = 1452

Example 6 : 198/88

Step 1:

$$\frac{88}{12} \quad 1 \ / \ 9 \ 8$$

88 is the divisor which has its base as 100 so we will have 2 digits in the remainder.

complement of 88 = 12
This is our new divisor.

Step 2:

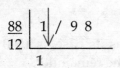

The first digit of the dividend is the first digit of the quotient.

Step 3:

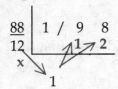

1×12 (new divisor) $= 12$

It is placed below the digits 98 of the dividend as shown.

Step 4:

$$\begin{array}{c|cc} 88 & 1/9 & 8 \\ 12 & 1 & 2 \\ \hline & 1/\mathbf{11} & 0 \end{array}$$

Since we are in the remainder side we will just add the digits in both the columns starting from the right.

$8 + 2 = {}_10$

$9 + 1 = 10 + 1 \ (carry) = 11$

So **110** becomes our remainder

Step 5:

$$\begin{array}{c|cc} 88 & 1 & /9 \ 8 \\ 12 & & 1 \ 2 \\ \hline & 1 & / \\ & & 1/1 \quad 0 \\ & & 1 \quad 2 \\ \hline & & 1 \ / 2 \ 2 \end{array}$$

Since 110 is more than the divisor 88 we have to divide it again.

Again we have 2 digits in the remainder. We bring down 1 as the quotient digit, multiply it with 12 (new divisor) and place it below 10, since we are again in the remainder we add both the columns

We get $Q = 1$ and $R = 22$

we add our new quotient to the previous quotient

Final $Q = 1 + 1 = 2$

$R = 22$

Answer : 198/88 gives **Q = 2 , R = 22**

Example 7 : 29137/919

Step 1:

$$\frac{919}{081} \Big| 2/9137$$

919 is the divisor which has its base as 1000 so we will have 4 digits in the remainder.

complement of 919 = 081
This is our new divisor.

Step 2:

$$\frac{919}{081} \Big| 2 \Big| /9137$$
$$\qquad 2$$

The first digit of the dividend is the first digit of the quotient.

Step 3:

$$\frac{919}{081} \Big| \begin{array}{l} 2\ 9/1\ 3\ 7 \\ \ \ 1\ 6\ 2 \end{array}$$
$$x \qquad 2$$

1×081 (new divisor) = 162

Place it below the digits 913 of the dividend as shown.

Step 4:

$$\frac{919}{081} \Big| \begin{array}{l} 2\ 9\ /1\ 3\ 7 \\ \ 1\ \ \ 6\ 2 \end{array}$$
$$\qquad 2\ _10\ /$$

We will add the digits in one column only, to get the next digit of the quotient.

$$9 + 1 = {}_10$$

This is the next digit in the quotient. But since it is a double digit we will write down 0 and take the other 1 as *carry over* to the previous digit of the quotient.

Step 5:

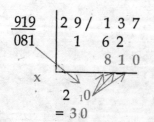

$$\frac{919}{081} \Big| \begin{array}{l} 2\ 9/\ 1\ 3\ 7 \\ \ \ 1\ \ \ 6\ 2 \\ \qquad \ 8\ 1\ 0 \end{array}$$
$$x$$
$$\quad 2\ _10$$
$$= 3\ 0$$

We multiply the whole number (10) with the new divisor and place it below the digits of the dividend.

$10 \times 081 = 810$

Place it below the digits 137 of the dividend as shown.

Our new quotient becomes 30

Step 6:

```
919 | 2 9 / 1 | 3 | 7
081 |   1 6 | 2 |
    |     8 | 1 | 0
```

30 / 15 6 7

Since we are in the remainder side we will just add the digits in all the columns starting from the right.

7 + 0 = 7

3 + 2 + 1 = 6

1 + 6 + 8 = 15

Remainder = 1567

This is bigger than the divisor so we divide it again.

Step 7:

```
919 | 2 9 / 1 3 7
081 |   1 6 2
    |     8 1 0
```

3 0 /

```
    1/ 5 | 6 | 7
     0 | 8 | 1
```

1 / 6 4 8

Since 1567 is more than the divisor 919 we have to divide it again.

Again we have 3 digits in the remainder. We bring down 1 as the quotient digit, multiply it with 081 (new divisor) and place it below 567, since we are again in the remainder we add all the columns.

We get Quotient = 1
and Remainder = 648

We add our new quotient to the previous quotient.

Final Q = 30 + 1 = 31
R = 648

Answer : 29137/919 gives **Q = 31 , R = 648**

Rules :

(1) We take the complement of the divisor and this becomes our new divisor.

(2) The first digit of the dividend is brought down as the first digit of the quotient.

(3) This digit is multiplied by the new divisor (complement of the original divisor) and written below the digits of the dividend. Write one digit in one column i.e. under each digit.

(4) Only the digits of the first column are added to get the next digit of the quotient.

(5) This quotient digit is again multiplied by the new divisor and written under the next digits of the dividend.

(6) The addition of digits in each column in each step, will give the new digit of the quotient

(7) We continue doing this till we have written a digit under all the digits of the remainder side.

(8) We then add the digits of the remainder side and get the remainder of the sum.

(9) If the remainder is bigger than the original divisor we again divide it by the new divisor,

- Add this quotient to the original quotient to get final quotient,
- The remainder of this part will be the final remainder.

Given below are a few more examples solving sums in straight steps

Example 8 : 2196452/89997   **Example 9** : 219356/988

```
89997 │ 2 1 / 9 6 4 5 2          988 │ 2 1 9 / 3 5 6
10003 │   2   0 0 0 6            012 │   0 2   4
      │       3 0 0 0 9              │     0   1 2
      └────────                     │         1 3 2
                                    └────────────
      2 3 / 1 / 2 6 5 2 1
              1 0 0 0 3          21 1 1 /   1 / 0 0 8
        ───────────             221           0 1 2
                                       ───────────
          1 / 3 6 5 2 4                 1 / 0 2 0
Answer : Q = 23 + 1 = 24        Answer : Q = 221 + 1 = 222
          R = 36524                         R = 20
```

Example 10 : 3242512/9958   **Example 11** : 2388964/9867

9958		3 2 4 / 2 5 1 2
0042		0 1 2 6
		0 0 8 4
		0 2 1 0

3 2 5 / 6 1 6 2

Answer : Q = 325
        R = 6162

9867		2 3 8 / 8 9 6 4
0133		0 2 6 6
		0 3 9 9
		1 3 3 0

23 10 / 2/ 0 8 8 4
240 /   0 2 6 6

2 / 1 1 5 0

Answer : Q = 240 + 2 = 242
       R = 1150

Exercise 10 .2 :

Attempt the following divisions:-

(1) 1211 / 779

(2) 1245 / 866

(3) 120034 / 9879

(4) 1248 / 987

(5) 24103 / 978

(6) 12131 / 7998

(7) 101008 / 898

(8) 11283 / 878

(9) 112131 / 8798

(10) 214121 / 9978

Answers

Exercise 10 .1

(1) Q-23, R-6	(2) Q-1366, R-7	(3) Q-133356, R-8	(4) Q-15, R-1
(5) Q-26, R-1	(6) Q-25, R-0	(7) Q-577, R-8	(8) Q-4668, R-2
(9) Q-3912, R-8	(10) Q-5704, R-6		

Exercise 10 .2

(1) Q-1, R-432	(2)Q- 1, R-379	(3) Q-12, R -1486	(4) Q-1, R-261
(5) Q-24, R-631	(6) Q-1, R-4133	(7) Q-112, R-432	(8) Q-12, R-747
(9) Q-12, R-6555	(10) Q-21, R-4583		

Glowing Star

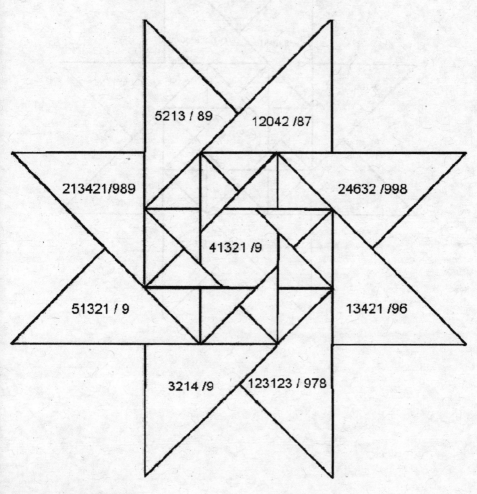

Solve the sums

If the digit in the tens place is :	colour the shape :
even	yellow
odd	orange

Within the figure:
- 5213 / 89
- 12042 /87
- 213421/989
- 24632 /998
- 41321 /9
- 51321 / 9
- 13421 /96
- 3214 /9
- 123123 / 978

Answers to Glowing Star

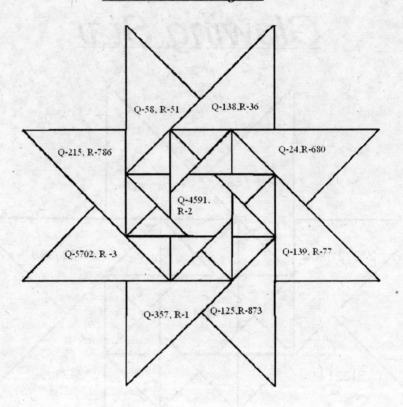

Chapter 11

DIVISION II

In the previous chapter we have studied the division by numbers less than the base, we will now discuss division by numbers more than the base. Here along with the "Nikhilam "we use the sutra

परावर्त्यं योजयेत्

"Parāvartya Yojayet"

"Transpose and Adjust"

We will follow almost the same method as in the previous division but with a little difference. In the previous cases when the divisor was less than the base our new divisor was its complement and a positive number. But here when the divisor is more than the base, we will still find its complement but we will now have a negative complement, so our new divisor will be negative. And then instead of adding the digits to get the next digit of our Quotient we will subtract them as the multiplication of the Quotient digit with the complement will be a negative number.

Let us understand it further with the help of a few examples:

<u>**Example 1**</u> : 1233/11

<u>Step 1:</u>

$$\begin{array}{c|c} 11 & 1\ 2\ 3\ /\ 3 \\ -1 & \underline{\qquad\qquad} \end{array}$$

We have 11 as our divisor, which is 1 more than the base 10.
So our new divisor will be -1.
We place our number as shown
As the base is 10 so we will have 1digit in the remainder.

Our new divisor will be - 1 or $\overline{1}$

Step 2:

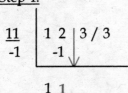

Here we bring down the first digit of the dividend as the first digit of the quotient as done in the previous divisions.

Step 3:

We multiply the first digit of the quotient by the new divisor and place it below the next digit (2) of the dividend as shown.

$$1 \times (-1) = -1 \text{ or } \overline{1}$$

Step 4:

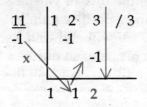

We will subtract the digits in the next column to get the next digit of the quotient.

$$2 - 1 = 1$$

Step 5:

We multiply the new digit of the quotient by the new divisor and place it below the next digit (3) of the dividend.
$$1 \times (-1) = -1$$

We will subtract the digits in the next column to get the next digit of the quotient.

$$3 - 1 = 2$$

Step 6:

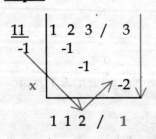

We multiply the new digit of the quotient by the new divisor and place it below the next digit (3) of the dividend or the remainder side.
$$2 \times (-1) = -2$$

We will subtract the digits in the remainder side to get the remainder.

$$3 - 2 = 1$$

Answer : 1233/11 gives **Q = 112 , R = 1**

Example 2 : 1346/12

Step 1:

$$\begin{array}{c|c} 12 & 1\ \ 3\ 4\ /\ 6 \\ -2 & \underline{\hspace{3cm}} \end{array}$$

12 is 2 more than the base 10
Our new divisor will be - 2

As the base is 10 we will have 1 digit in the remainder side.

Step 2:

$$\begin{array}{c|c} 12 & 1\ |\ 3\ 4\ /\ 6 \\ -2 & \downarrow \\ & 1 \end{array}$$

The first digit of the dividend is the first digit of the quotient.

Step 3:

$$\begin{array}{c|c} 12 & 1\ \ 3\ 4\ /\ \ 6 \\ -2 & \ \ -2 \\ x & 1 \end{array}$$

Multiply the first digit of the quotient by the new divisor and place it below the next digit (3) of the dividend as shown.

$$1 \times (-2) = -2 \text{ or } \overline{2}$$

Step 4:

$$\begin{array}{c|c} 12 & 1\ 3\ |\ 4\ /\ 6 \\ -2 & \ \ -2 \\ & 1\ \ 1 \end{array}$$

Subtract the digits in the next column to get the next digit of the quotient.

$$3 - 2 = 1$$

Step 5:

$$\begin{array}{c|c} 12 & 1\ 3\ 4\ \ |\ /\ 6 \\ -2 & \ \ -2\ -2 \\ x & 1\ 1\ \ 2 \end{array}$$

Now
$$1 \times (-2) = -2$$
We will subtract the digits in the next column to get the next digit of the quotient.

$$4 - 2 = 2$$

Step 6:

$$\begin{array}{c|c} 12 & 1\ 3\ 4\ /\ \ 6 \\ -2 & \ \ -2\ -2\ \ \ -4 \\ x & 1\ 1\ 2\ /\ 2 \end{array}$$

Now
$$2 \times (-2) = -4$$
We will subtract the digits in the remainder side to get the remainder.
$$6 - 4 = 2$$

Answer : 1346/12 gives **Q = 112 , R = 2**

Example 3 : 2561/102

Step 1:

102 is 02 more than the base 100
As the base is 100 we have two digits in the remainder.

$$\underline{\begin{array}{c} 102 \\ -0-2 \end{array}} \bigg|\ 25\,/\,6\,1$$

The new divisor will be - 02 written as - 0 - 2

It is important to keep in mind the base of the divisor to find the new divisor. Otherwise it will be as if it is not taken correctly and the whole answer will go wrong.

Here we have to take it as – 0 - 2 and not – 2

The rest of the steps are followed as the previous examples.

Step 2 :

$$\begin{array}{c|ccc} 102 & 2\ 5 & /\ 6 & 1 \\ -0\text{-}2 & -0 & -4 & \\ & & -1 & -0 \\ \hline & 2\ 5 & /\ 1 & 1 \end{array}$$

(1) 2 is brought down as first digit of the Q
(2) 2 × (- 02) = - 04, (write - 0 -4 below 56)
(3) 5 – 0 = 5, next digit of Q
(4) 5 × (- 02) = -10, (write - 1 - 0 below 61)
(5) We are in the remainder side, so we subtract from right side
     1 – 0 = 1 ,
     6 – 4 –1 = 1 this is the remainder

(Notice that we have written -04 as –0 - 4 and other two digit numbers in the same pattern in the sum for calculation purposes as we are not subtracting the whole number but each digit of the number in different steps and in -04 both the digit are negative so we write it as -0 -4 .)

Answer : 2561/102 gives **Q = 25 , R = 11**

Example 4 : 10569/103

Step 1:

$$\underline{103} \bigm| 105 / 69$$
-0 -3

103 is 3 more than the base 100 but we will write it as 03 as 100 has 2 zeros.

-0 -3 is the new divisor
As base is 100 we will have 2 digits in the remainder side.

Step 2:

$$\begin{array}{c|c} 103 & 1\ 0\ 5\ /\ 6\ 9 \\ -0\text{-}3 & \ -0\ \text{-}3 \\ & \quad -0\ \ \text{-}0 \\ & \qquad -0\ \ \text{-}6 \\ \hline & 1\ 0\ 2\ /\ 6\ 3 \end{array}$$

(1) 1 is brought down as first digit of Q.
(2) $1 \times (-03) = -03$,(write - 0 - 3 below 05)
(3) Subtract digit of one column : 0 - 0 = 0
(4) $0 \times (-03) = -00$ (write - 0 - 0 below 56)
(5) Subtract digits of one column :
    5 – 3 – 0 = 2, next digit of Q
(6) $2 \times (-03) = -06$, (write - 0 - 6 below 69)
(7) Subtract the remainder digits from the right :
    9 – 6 = 3
    6 – 0 – 0 = 6
    Remainder = 63

Answer : 10569/103 gives **Q = 102 , R = 63**

Example 5 : 16453/102

Step 1:

$$\underline{102} \bigm| 164 / 53$$
- 0- 2

102 is 2 more than the base 100 but we will write it as 02 as 100 has 2 zeros.

- 0 - 2 is the new divisor
As base is 100 we will have 2 digits in the remainder side.

Step 2:

$$\begin{array}{c|c} 102 & 1\ 6\ 4\ /\ 5\ 3 \\ -0\text{-}2 & \ -0\ \text{-}2 \\ & \quad -1\ \ \text{-}2 \\ & \qquad -0\ \ \text{-}2 \\ \hline & 1\ 6\ 1\ /\ 3\ 1 \end{array}$$

(1) 1 is brought down as first digit of Q
(2) $1 \times (-02) = -02$,(write - 0 - 2 below 64)
(3) Subtract digits of one column : 6 - 0 = 6
(4) $6 \times (-02) = -12$ (write - 1 - 2 below 45)
(5) Subtract digits of one column :
    4 – 2 – 1 = 1, next digit of Q
(6) $1 \times (-02) = 02$, (write 02 below 53)

(7) Subtract the remainder digits from the right:

$$3 - 2 = 1$$
$$5 - 2 - 0 = 3$$
$$\text{Remainder} = 31$$

Answer : 16453/102 gives **Q = 161, R = 31**

Example 6 : 23689/112

Step 1:

112	2 3 6 / 8 9
-1-2	

112 is 12 more than the base 100

- 1 - 2 is the new divisor
As base is 100 we will have 2 digits in the remainder.

Step 2:

```
112 | 2  3   6 / 8   9
-1-2|   -2  -4
    |       -1  -2
    |           -1  -2
    |_____
      2  1   1 / 5   7
```

(1) 2 is brought down as first digit of Q.
(2) 2 ×(-12) = - 24, (write – 2 - 4 below 36)
    Subtract digits of one column : 3 - 2 = 1
(3) 1 × (- 12) = - 12 (write - 1 - 2 below 68)
(4) Subtract digits of one column :
    6 – 4 – 1 = 1, next digit of Q
(6) 1 × (- 12) = - 12, (write - 1 - 2 below 89)
(7) Subtract the remainder digits :
    9 – 2 = 7
    8 – 2 – 1 = 5
    Remainder = 57

Answer : 23689/112 gives **Q = 211, R = 57**

Rules :

(1) The excess of the divisor is made the new divisor with a negative sign.

(2) The first digit of the dividend is brought down as the first digit of the quotient.

(3) This digit is multiplied by the new (negative) divisor.

(4) The digits are placed under the digits of the dividend one digit in each column i.e. under each digit of the dividend.

(5) Only the digits of one column are subtracted to get the next digit of the quotient.

(6) This is again multiplied by the new divisor and written under the next digits of the dividend.

(7) The subtraction of digits in each column in each step will give the next digit of the quotient.

(8) We continue doing this till we have written a digit under all the digits in the remainder side.

(9) We then subtract the digits in the remainder side starting from the right and get the remainder of the sum.

Exercise 11 .1 :

Attempt the following divisions :-

(1) 1479/12

(2) 15649/12

(3) 29694/14

(4) 24365/11

(5) 15921/106

(6) 312749/104

(7) 146785/121

(8) 23284/1103

(9) 25483/1203

(10) 25953/123

Pentagonal Prism

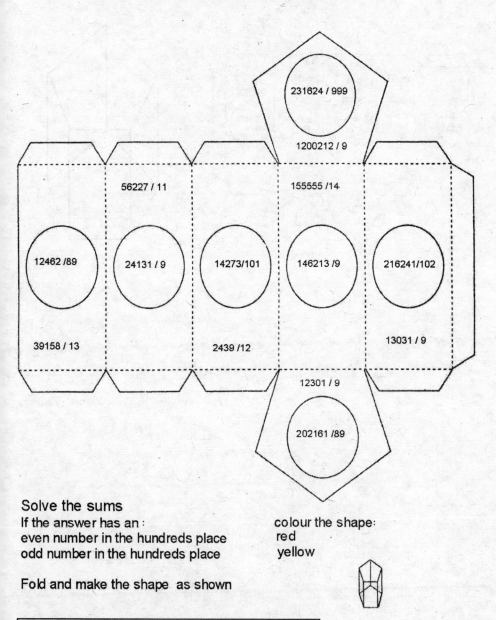

231624 / 999

1200212 / 9

56227 / 11

155555 /14

12462 /89

24131 / 9

14273/101

146213 /9

216241/102

39158 / 13

2439 /12

13031 / 9

12301 / 9

202161 /89

Solve the sums
If the answer has an :
even number in the hundreds place
odd number in the hundreds place

colour the shape:
red
yellow

Fold and make the shape as shown

Cut out the pattern along the outer solid lines and fold
along the dotted lines. Tape each flap to the under side
of the pentagon or rectangle shape.

<u>Answers to Pentagonal Prism</u>   <u>Final Shape</u>

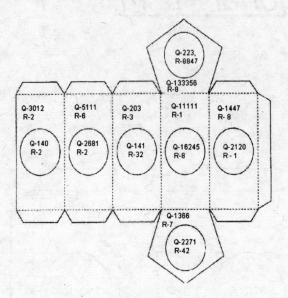

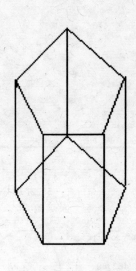

Special Cases

There are some peculiar cases in this type of division, since we are subtracting in each step of the division we can get a negative number in the remainder side or even a negative digit in the quotient. Let us now study a few examples for this type and understand the method to be followed.

First we will take a few examples when we get a negative remainder.

Example 1 : 123456/1123

Step 1:

1123	1 2 3 / 4 5 6
-1-2-3	

1123 is 123 more than the base 1000.
- 1- 2- 3 is the new divisor

As base is 1000 we will have 3 digits in the remainder.

Step 2:

1123	1 2 3 / 4 5 6
-1-2-3	-1 -2 - 3
	-1 -2 -3
	-0 -0 -0
	1 1 0 /-1 2 6

(1) 1 is brought down as first digit of Q.
(2) $1 \times (-123) = -123$, ($-1-2-3$ in each column)
(3) Subtract digits of one column : $2 - 1 = 1$
(4) $1 \times (-123) = -123$ ($-1-2-3$ in each column)
(5) Subtract digits of one column :
$3 - 2 - 1 = 0$, next digit of Q
(6) $0 \times (-123) = 000$, ($-0-0-0$ in each column).
(7) Subtract the remainder digits from the right
$6 - 0 = 6$
$5 - 3 - 0 = 2$
$4 - 3 - 2 = -1$

Digits of remainder are $= -1 + 2 + 6$

When we take their place value we have
$- 100, + 20$ and $+ 6$ as the remainder,
So remainder $= -100 +20 +6 = -74$

Since we have negative remainders we carry over 1 from the quotient side to the remainder side i.e. take 1123 over to the remainder side and subtract the negative remainder from this.
Final R $= 1123 - 74 = 1049$
Q $= 110 - 1 = 109$
Answer : 13456/1123 gives **Q = 109 , R = 1049**

Example 2 : 11211/1012

Step 1:

$$\begin{array}{c|c} 1012 & 11\,/\,211 \\ \hline -0\text{-}1\text{-}2 & \end{array}$$

1012 is 012 more than the base 1000
- 0 - 1 - 2 is the new divisor.

As base is 1000 we have 3 digits in the remainder.

Step 2:

$$\begin{array}{c|c} 1012 & 1\ \ 1\,/\,2\ \ 1\ \ 1 \\ -0\text{-}1\text{-}2 & \ \ -0\ -1\ -2 \\ & \qquad -0\ -1\ -2 \\ \hline & 11\,/\ \ 1\ -2\ -1 \end{array}$$

(1) 1 is brought down as the first digit of the Q
(2) $1 \times (-012) = -012$, $(-0$ -1 -2 in each column)
(3) Subtract digits of one column :
$\quad 1 - 0 = 1$, next digit of Q
(4) $1 \times (-012) = -012$, $(-0$ -1 -2 in each column)
(5) Subtract the digits of the remainder :
$\quad 1 - 2 = -1$
$\quad 1 - 2 - 1 = -2$
$\quad 2 - 1 - 0 = 1$

Digits of the remainder are = 1 - 2 -1
When we take their place value,
R = +100 - 20 - 1 = 79 (this is positive)
Final   R = 79
$\qquad$ Q = 11

Answer : 11211/1012 gives **Q = 11, R = 79**

Example 3 : 13358/1133

Step 1:

$$\begin{array}{c|c} 1133 & 13358 \\ \hline -1\text{-}3\text{-}3 & \end{array}$$

1133 is 133 more than the base 1000.

New divisor is - 1 - 3 - 3
As base is 1000 we have 3 digits in the remainder.

Step 2:

$$\begin{array}{c|c} 1133 & 13\,/\,3\ \ 5\ \ 8 \\ -1\text{-}3\text{-}3 & \ -1\ \ -3\ -3 \\ & \qquad -2\ -6\ -6 \\ \hline & 1\ 2\,/\,-2\ -4\ +2 \end{array}$$

(1) 1 is brought down as the first digit of the Q
(2) $1 \times (-133) = -133$, $(-1$ -3 -3 in each column)
(3) Subtract digits of one column :
$\qquad 3 - 1 = 2$, next digit of Q
(4) $2 \times (-133) = -266$, $(-2$ -6 -6 in each column)
(5) Subtract the digits of the remainder :
$\qquad 8 - 6 = 2$

$$5 - 3 - 6 = -4$$
$$3 - 3 - 2 = -2$$

Digits of the remainder are $= -2 - 4 + 2$

When we take their place value,

$R = -200 - 40 + 2$

$R = -238$

Since remainder is negative we carry 1 from quotient to remainder side.

(we carry 1 dividend : 1133)

Final $R = 1133 - 238 = 895$

        $Q = 12 - 1 = 11$

Answer : 13358/1133 gives $Q = 11$, $R = 895$

Now let us take an example if we get a negative digit in the middle of the Quotient.

Example 4 : 8623/112

Step 1:

```
112 | 8 6 2 3
-1-2 |_____
```

112 is 12 more than the base 100

New divisor is $-1 - 2$

As base is 100 we have 2 digits in the remainder.

Step 2:

```
112  | 8 6 / 2  3
-1-2 | -9  -6
     |      +3 +6
     |_____
       8 - 3/ -1 + 9
```

(1) 8 is brought down as the first digit of the Q

(2) $8 \times (-12) = -96$, $(-9 - 6$ in each column$)$

(3) Subtract digits of one column :

    $6 - 9 = -3$, next digit of Q

(4) $-3 \times (-12) = +36$, $(+3 + 6$ in each column$)$

(5) Subtract / add the digits of the remainder :

    $3 + 6 = 9$

    $2 - 6 + 3 = -1$

    $R = -1 + 9$

    $\cdot = -10 + 9$

    $R = -1$

In the quotient we have 8 – 3
When we take their place value it is
$$80 – 3 = 77$$
$$Q = 77$$
Since remainder is negative we carry 1 from
quotient to remainder side.

$$\text{Final } Q = 77 – 1 = 76$$
$$R = 112 – 1 = 111$$

Answer : 8623/112 gives **Q = 76, R = 111**

Example 5 : 438603/1121

Step 1:   1121 is 121 more than the base 1000.
                             New divisor is **- 1 - 2 - 1**

$$\begin{array}{c|c} 1121 & 438 / 603 \\ \text{-1-2-1} & \end{array}$$

As base is 1000 we have 3 digits in the
remainder.

Step 2:   (1) **4** is brought down as the first digit of the
                                 Q

$$\begin{array}{c|cc} 1121 & 4\ 3\ 8\ /\ 6\ \ 0\ \ 3 \\ \text{-1-2-1} & -4\ -8\ \ -4 \\ & \quad +1\ +2\ +1 \\ & \qquad\quad -1\ -2\ -1 \\ \hline & 4\ -11\ /\ 3\ -1\ +2 \end{array}$$

(2) 4× (- 121) =- **484** , (-4 -8 -4 in each column)
(3) Subtract digits of one column :
        3 - 4 = -1, next digit of Q
(4) - 1 × (- 121) =+**121** , (+1+2+1 in each
        column)
(5) Subtract digits of one column :
        8 - 8 +1 = 1, next digit of Q
(6) 1 x (- 121) =- **121** , (-1-2-1 in each column)
(7) Subtract/add the digits of the remainder :
        3 - 1 =2
        0 +1 - 2 = - 1
        6 - 4 + 2 - 1 =3

$$R = 3\ -1 + 2$$
$$= 300\ - 10 + 2 \text{ (place value)}$$
$$\text{Final }\ R = 292$$

In the Quotient we have 4 -1+1 which
actually means
400 – 10 +1 =
Final   Q = 391

*When we get a negative digit as part of the
quotient then we get a positive number in the next
step of calculations as in step 3.*

Answer : 438603/1121 gives **Q= 391, R = 292**

Rules :

(1) The excess of the divisor is made the new divisor with a negative sign.

(2) The first digit of the dividend is brought down as the first digit of the quotient.

(3) This digit is multiplied by the new (negative) divisor.

(4) The digits are placed under the digits of the dividend one digit in each column i.e. under each digit.

(5) Only the digits of one column are subtracted to get the next digit of the quotient.

(6) This is again multiplied by the new divisor and written under the next digits of the dividend.

(7) The subtraction of digits in each column in each step will give the next digit of the quotient.

(8) We continue doing this till we reach all the digits in the remainder side.

(9) We then subtract the digits in the remainder side starting from the right and get the remainder of the sum.

(10) If any digit in the remainder is negative then the actual remainder is calculated and if it is positive this becomes our final remainder but if the remainder is negative then one (dividend) is carried from the quotient to the remainder side.

   • One is subtracted from the quotient to get the final quotient.

   • The negative remainder is subtracted from the dividend to get the final remainder.

(11) If any digit in the quotient is negative then we calculate the actual quotient by taking the place value of each digit and subtracting the negative parts from the positive parts and since the first digit of the quotient will always be positive, we will get a positive quotient as the final quotient.

Given below are a few more examples solving the sums in straight steps:

Example 6 : 137712/102

```
102 | 1  3   7   7 / 1  2
-0-2|   - 0 - 2
    |      - 0 - 6
    |         - 1   - 0
    |            - 0 - 0
    |_____
      1  3  5  0 / 1  2
```

Answer : Q = 1350, R = 12

Example 7 : 24841/112

```
112 | 2  4  8 / 4   1
-1-2|   - 2 - 4
    |      - 2 - 4
    |         - 2  - 4
    |_____
      2  2  2 / - 2  - 3
```

Q = 222 – 1 = 221
R = 112 – 23 = 89

Answer : Q = 221, R = 89

Example 8 : 27614/121

```
121 | 2  7  6 / 1  4
-2-1|   - 4 - 2
    |     - 6  - 3
    |            4  2
    |_____
      2  3 - 2 / 2  6
```

Q = 230 – 2 = 228
R = 26

Answer : **Q = 228, R = 26**

Example 9 : 21841/117

```
117 | 2  1  8 / 4  1
-1-7|   - 3 - 4
    |        3   4
    |          - 11  - 9
    |_____
      2 - 2  7 / - 3  - 8
```

R = - 3 -8 = -38
Q = 207 – 20 = 187
      187 – 1 = 186
R = 117 – 38 = 79

Answer : **Q = 186, R = 79**

Exercise 11 .2 :

Attempt the following divisions :-

(1) 21787/111

(2) 10269/103

(3) 15542/114

(4) 25181/1203

(5) 7685/112

(6) 25728/1112

(7) 12349/1133

(8) 13905/113

(9) 13999/112

(10) 104135/1121

Answers

Exercise 11 .1

(1) Q-123, R -3   (2) Q-1304, R-1   (3) Q-2121, R-0   (4) Q-2215, R-0

(5) Q-150, R-21   (6) Q-3007, R-21   (7) Q-1213, R-12   (8) Q-21, R-121

(9) Q-21, R-220   (10) Q-211, R-0

Exercice 11 .2

(1) Q-196, R-31   (2) Q-99, R-72   (3) Q-136, R-38   (4) Q-20, R-1121

(5) Q-68, R-69   (6) Q-23, R-152   (7) Q-10, R-1019   (8) Q-123, R-6

(9) Q-124, R-111   (10) Q-92, R-1003

Chapter 12

SQUARES

Before we study squares in details let us first learn how to find squares of specific numbers. In this chapter we will deal with squares numbers ending in a particular digit. If our given number falls in any of the following categories then the working out the square of that number becomes very easy. Like numbers ending in 5, 1, 4, 6, big number that are close to bases, etc.

Squares of Numbers ending in 5

To find the square of any number ending in five we use the sutra

एकाधिकेन पूर्वेन

"Ekādhikena Pūrvena"

"By one more than the one before"

Let us take a few examples and understand its use

Example 1: 35^2

35^2 The number 35 ends with 5. We get the answer in two parts, Right Hand Side (R.H.S.) and Left Hand Side (L.H.S.)

Step 1:

$3\,5^2$

$\downarrow$

$/\ 25$

R.H.S. :
As our number ends in 5 and the square of 5 is 25. So right side of our answer is

25

Step 2:

$3\,5^2$

$\times 4$

$\downarrow$

$12/\ 25$

L.H.S. :
Previous digit of 5 in 35 is 3 and one more than 3 is 4, So

$3 \times 4 = 12$

In other words multiply the previous digit 3 by its next number 4.

Answer : $35^2 = 1225$

Example 2 : 75^2

$$7\ 5^2$$

x 8

$$= 56 \ / \ 25$$

We get the answer in two parts.

<u>R.H.S.</u> : The number ends in 5 and
$5^2 = 25$

<u>L.H.S.</u> : Multiply the previous digit 7 by its next number 8.
$$7 \times 8 = 56$$

Answer : $75^2 = 5625$

Example 3: 125^2

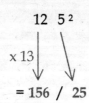

$$12\ 5^2$$

x 13

$$= 156 \ / \ 25$$

We get the answer in two parts.

<u>R.H.S.</u> : 25 (as the number ends in 5)

<u>L.H.S.</u> : Previous digit multiplied by its next number. Here we have 2 digits prior to 5, we take both the digits together and multiply them by the next number.
$$12 \times 13 = 156$$

Answer : $125^2 = 15625$

<u>Rules :</u>

(1) We get our answer in two parts.

(2) Right hand side will always be 25 as the number ends in 5, so its square will end in 25.

(3) The previous part of the number multiplied by its next number will form the left hand side of the answer.

(4) If we have a bigger number then all the other digits except 5 are taken as the previous part of the number and they are multiplied by their next number to get the left part of the answer.

Given below are a few more examples solving sums in straight steps:-

$85^2 = (8 \times 9) / 25 = 7225$

$135^2 = (13 \times 14) / 25 = 18225$

$265^2 = (26 \times 27) / 25 = 70225$

$1215^2 = (121 \times 122) / 25 = 1476225$

When we compare both the methods we find that the Vedic method makes our calculations very short.

135^2

Usual method	Vedic method
135 x 135 ---------- 675 4050 13225 ------------- 18225 -------------	$(13 \times 14) / 25$ $= 18225$

Exercise 12 .1 :

Find the following squares mentally :-

(1) 25^2 (6) 205^2

(2) 65^2 (7) 115^2

(3) 45^2 (8) 225^2

(4) 85^2 (9) 505^2

(5) 105^2 (10) 1005^2

We get a very interesting multiplication of numbers where the digits in the units place add up to 10 and the digits in the tens place are the same in both the numbers by taking the square of numbers ending with 5 a little further, along with the use of the sutra

अन्त्ययोर्दशकेऽपि

"Antyayordaśake'pi"

"Last totalling to ten"

Let us try and understand it further with the help of a few examples:

Example 1 : 43 x 47

Step 1:

4 3 x 4 7

/ 21

Here we observe that the units of both the numbers add up to 10 (3 + 7 = 10) and the tens are the same (4)
We get the answer in two parts.

R.H.S. : Multiplication of both the unit digits
7 x 3 = 21

Step 2:

$$4\ 3 \times 4\ 7$$
$$\times 5$$
$$\downarrow$$
$$20\ /\ 21$$

L.H.S. : The tens digit (common previous digit) multiplied by its next number.

$$4 \times 5 = 20$$

Answer: $43 \times 47 = \mathbf{2021}$

Example 2 : 22 x 28

Step 1:

$$2\ 2 \times 2\ 8$$
$$\times$$
$$/\ 16$$

Here we observe that the units of both the numbers add up to 10 and the tens are the same (2).
We get answer in two parts.

R.H.S. :
Multiplication of both the unit digits
$$2 \times 8 = 16$$

Step 2:

$$2\ 2 \times 2\ 8$$
$$\times 3$$
$$\downarrow$$
$$6\ /\ 16$$

L.H.S. :
2 (The common previous digit) multiplied by its next number (3).

$$2 \times 3 = 6$$

Answer : $22 \times 28 = \mathbf{616}$

Example 3 : 61 x 69

$$6\ 1 \times 6\ 9$$
$$\times 7$$
$$42\ /\ 09$$

R.H.S. : $1 \times 9 = 09$

(In this method of multiplication, 100 is taken as the base so we should always have 2 digits on the right side.)

L.H.S. : 6×7 (its next numbers) $= 42$

Answer : $61 \times 69 = \mathbf{4209}$

Example 4 : 112 x 118

R.H.S. : $2 \times 8 = 16$

11 2 x 11 8

L.H.S. : $11 \times 12 = 132$

x 12

132 / 16

Answer : $112 \times 118 = 13216$

Rules :

(1) Used for multiplication in which the units digit of both the numbers add up to 10 and the tens digits are the same, or the all the digits prior to the units digit are the same.

(2) The right side of the answer is the multiplication of the units' digits.

(3) The left side of the answer is the multiplication of digit(s) prior to the unit digit multiplied by the next number.

Given below are a few more examples solving sums in straight steps:

Example 5 :

92 x 98
= 90 / 16

Answer : 9016

Example 6 :

74 x 76
= 56 /24

Answer : 5624

Example 7 :

121 x 129
= 156 /09

Answer : 15609

Example 8 :

397 x 393
= 1560/21

Answer : 156021

When we compare both the methods we find that the Vedic method makes our calculations very short.

127 x 123

<u>Usual method</u>

```
    127
  x 123
-----------
    381
   2540
  12700
-----------
  15621
-----------
```

<u>Vedic method</u>

127 x 123
156/21
= 15621

Exercise 12 .2 :

Try and solve these sums mentally:-

(1) 62 x 68

(2) 73 x 77

(3) 81 x 89

(4) 44 x 46

(5) 113 x 117

(6) 214 x 216

(7) 317 x 313

(8) 441 x 449

(9) 1112 x 1118

(10) 1006 x 1004

Square of Numbers ending in 1

This specific rule is followed to find the square of a number ending in 1. We get the answer in 3 parts as explained in the examples below.

Example 1 : 31^2

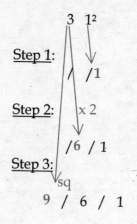

We get the answer in three parts.

Step 1:   R.H.S. : The answer is always 1.as $1^2 = 1$

Step 2:   Middle : Double the previous part (3) of the number.

$$3 \times 2 = 6$$

Step 3:   L.H.S. : Square of the previous part (3) of the number.

$$3^2 = 9$$

Answer : $31^2 = 961$

Example 2 : 41^2

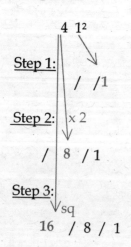

We get the answer in three parts.

Step 1:   R.H.S. :1

Step 2:   Middle : Double the previous part (4) of the number.

$$4 \times 2 = 8$$

Step 3:   L.H.S. : Square of the previous part of the number.

$$4^2 = 16$$

Answer : $41^2 = 1681$

Example 3 : 61^2

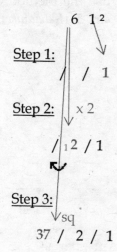

Step 1:

Step 2: $\times 2$

$/\,_12\,/\,1$

Step 3:

sq

$37\,/\,2\,/\,1$

We get the answer in three parts.

R.H.S. : 1

Middle : 6 (previous part) x 2 $=_12$

We write single digit at a time so, we put down 2 and *carry over* 1 to next step.

L.H.S. : 6^2 (previous part) = 36
     36 +1 (*carry*)   = 37

Answer : 61^2 = **3721**

Example 4 : 121^2

12 1²

146 $/\,_24\,/\,1$

We get the answer in three parts.
R.H.S. :1

Middle : 12 (previous part) x 2 $=_24$
     We put down 4 and *carry over* 2..

L.H.S. : 12^2 (previous part) = 144
     144 +2 (*carry*) =146

Answer : 121^2 = **14641**

Example 5 : 351^2

35 1²

$/\,_70\,/1$

We get the answer in three parts.
R.H.S. : 1

Middle : 35 (previous part) x 2 $=_70$
We put down 0 and *carry over* 7

1232 / 0 / 1

L.H.S. : 35^2

As 35 ends in 5 so we can use the method of squares of number ending in 5 (*Ekhadhikena Purvena*).

$$35^2 = (3 \times 4) / 25 = 1225$$

$$1225 + 7 \ (carry) = 1232$$

Answer : $351^2 = 123201$

Rules :

(1) When a number ends in 1 then we get our answer in 3 parts.

(2) Right hand side of the answer is always 1 as the number ends in 1 and square of 1 is always 1.

(3) Middle part is the double of the digit(s) before 1. We put down only a single digit, but if we have more than one digit then we take the other digits as carry over to the next step.

(4) Left hand side is the square of the previous part of the number other than 1.We add the carry over, if any, to this part from the middle part and write down the answer.

Given below are a few more examples solving sums in straight steps:

Example 6 : 81^2

$$81^2 = 64 / {}_16 / 1$$
$$= 6561$$

Answer : **6561**

Example 7 : 131^2

$$131^2 = 169 / {}_26 / 1$$
$$= 17161$$

Answer : **17161**

Example 8 : 851²

851² = 7225 / ₁₇0 /1
     = 724201

Answer : **724201**

Example 9 : 711²

711² = (49/₁4/1) / ₁₄2 / 1
     = 5041 / ₁₄2 / 1
     = 505521

Answer : **505521**

If we compare both the methods we see the Vedic method makes our calculations very short.
951²

Usual method

   951
x 951

   951
 47550
855900

904401

Vedic method

9025 /190 / 1

= 904401

Exercise 12 . 3 :

Find the squares of the following numbers mentally :-

(1) 21²

(2) 51²

(3) 71²

(4) 91²

(5) 101²

(6) 111²

(7) 151²

(8) 251²

(9) 211²

(10) 1201²

Squares of Numbers ending with 4

When any number ends in 4 then we follow a specific method to find its square.

Example 1 : 34^2

Step 1:

34^2

$= 35^2 - (34 + 35)$

We know 34 is close to 35
First we find the square of its nearest number ending with 5.

$35^2 = (3 \times 4) / 25 = 1225$
(use *Ekadhikena Purvena* to find the square of numbers ending with 5)

Step 2:

$34^2 = 1156$

We then subtract, the sum of the given number and its nearest number ending with 5
i.e. $(34 + 35) = 69$
from the square of the number ending with 5.
$= 1225 - 69$
$= 1156$

Answer : $34^2 = \mathbf{1156}$

Example 2 : 94^2

Step 1:

94^2

$= 95^2 - (94 + 95)$

We know 94 is close to 95
We find the square of its nearest number ending with 5
$95^2 = (9 \times 10) / 25 = 9025$
(use *Ekadhikena Purvena*)

Step 2:

$94^2 = 8836$

We then subtract, the sum of the number and its nearest number ending with 5
i.e. $(94 + 95) = 189$
from the square of the number ending with 5.
$= 9025 - 189$
$= 8836$

Answer : $94^2 = 8836$

Example 3 : 624 ^{2}

624 ^{2} = 389376

We know 624 is close to 625
(1) 625^2 = (62 x 63) / 25 = 390625
(2) (624 + 625) = 1249
    390625 – 1249 = 389376

Answer : 624 ^{2} = 389376

Rules :

(1) We find the square of the previous number of the given number ending in 5, closest to the number (its next number).

(2) Subtract the sum of the given number and the number ending in 5 from the square of the number ending in 5.

Given below are a few more examples solving sums in straight steps:

Example 4 : 84 ^{2}

84 ^{2} = 85^2– (84+85)
   = 7225 –169
   = 7056

Answer : 7056

Example 5 : 114 ^{2}

114 ^{2} = 115^2– (114+ 115)
    = 13225 –229
    = 12996

Answer: 12996

Example 6 : 244 ^{2}

244 ^{2} = 245^2– (244+ 245)
   = 60025 –489
   = 59536

Answer : 59536

Example 8 : 374 ^{2}

374 ^{2} = 375^2– (374+ 375)
    = 140625 – 749
    = 139876

Answer : 139876

When we compare both the methods we find that the Vedic method makes our calculations very short.

244^2

Usual method	Vedic method
244	$245^2 - (244 + 245)$
x 244	$= 60025 - 489$
------------	$= 59536$
976	
9760	
48800	

59536	

Exercise 12 .4 :

Find the squares of the following numbers :-

(1) 44^2 (6) 154^2

(2) 64^2 (7) 294^2

(3) 74^2 (8) 254^2

(4) 94^2 (9) 544^2

(5) 134^2 (10) 954^2

Squares of any Number ending with 6

For any number that ends in 6 we follow the same method as in numbers ending in 4 but with a little difference.

Let us take a few examples to understand it.

Example 1 : 36^2

Step 1:

36^2

$= 35^2 + (35 + 36)$

We know 36 is close to 35
We find the square of its nearest number ending with 5.

$35^2 = (3 \times 4) / 25 = 1225$
(use *Ekadhikena Purvena*)

Step 2:

$36^2 = 1296$

We then add the sum of the number and its nearest number ending with 5
i.e. $(35 + 36) = 71$
to the square of the number ending with 5.
$= 1225 + 71$
$= 1296$

Answer: $36^2 = 1296$

Example 2 : 76^2

Step 1:

76^2

$= 75^2 + (75 + 76)$

We know 76 is close to 75
We find the square of its nearest number ending with 5.

$75^2 = (7 \times 8) / 25 = 5625$
(use *Ekadhikena Purvena*)

Step 2:

$76^2 = 5776$

We then add the sum of the number and its nearest number ending with 5
i.e. $(75 + 76) = 151$
to the square of the number ending with 5.
$= 5625 + 151$
$= 5776$

Answer : $76^2 = 5776$

Example 3 : 156^2

<table>
<tr><td>$156^2 = 24336$</td><td>We know 156 is close to 155
(1) $155^2 = (15 \times 16) / 25 = 24025$
(2) $(155 + 156) = 311$
$\quad\quad 24025 + 311 = 24336$</td></tr>
</table>

Answer : $156^2 = 24336$

Rules :

(1) We find the square of the previous number of the given number ending in 5, closest to the number (its next number).

(2) Add the sum of the given number and the number ending in 5 to the square of the number ending in 5.

Given below are a few more examples solving sums in straight steps :

Example 4 : 76^2

$76^2 = 75^2 + (75 + 76)$
$\quad\quad = 5625 + 151$
$\quad\quad = 5776$

Answer : 5776

Example 5 : 156^2

$156^2 = 155^2 + (155 + 156)$
$\quad\quad = 24025 + 311$
$\quad\quad = 24336$

Answer : 24336

Example 6 : 246^2

$246^2 = 245^2 + (245 + 246)$
$\quad\quad = 60025 + 491$
$\quad\quad = 60516$

Answer : 60516

Example 7 : 816^2

$816^2 = 815^2 + (815 + 816)$
$\quad\quad = 664225 + 1631$
$\quad\quad = 665856$

Answer : 665856

When we compare both the methods we find that the Vedic method makes our calculations very short.

246^2

Usual method	Vedic method
	$= 245^2 + (245 + 246)$
246	$= 60025 + 491$
x 246	$= 60516$

1476	
9840	
49200	

60516	

Exercise 12 .5 :

Find the squares of the following numbers :-

(1) 46^2 (6) 176^2

(2) 66^2 (7) 286^2

(3) 86^2 (8) 326^2

(4) 96^2 (9) 566^2

(5) 126^2 (10) 956^2

Square of Number closer to the Bases

For squares of numbers close to the base we use a sub sutra of the "Nikhilam" sutra

यावदूनं तावदूनीकृत्य वर्गं च योजयेत्

"Yāvadūnam Tāvadūnīkṛtya Vargañca Yojayet"

*"Whatever the extent of its deficiency, lessen by that amount,
and set of the square of that deficiency"*

The meaning of the above will be clearer with a few examples. We will first take examples to find square of numbers more than the base.

Example 1 : 103^2

<u>Step 1:</u>	Base for 103 is 100 and it is 3 more than the base.
	We get the answer in 2 parts.
/ 09	
	<u>R.H.S.</u> : We take the square of the complements
	$3^2 = 09$
	since 100 is the base we write it as **09**
<u>Step 2:</u>	<u>L.H.S.</u> : 103 is 3 more than the base (100)
	We add the excess to the number
106/ 09	$103 + 3 = 106$

Answer : $103^2 = 10609$

Example 2 : 112^2

<u>Step 1:</u>	Base for 112 is 100 and it is 12 more than the base.
	We get the answer in 2 parts.
/ ₁44	
	<u>R.H.S.</u> : We take the square of the complements
	$12^2 = {}_1 44$

But 100 is the base so 2 digits on right side. So we write

44 and *carry over* 1 to right side

<u>Step 2:</u>

125 / 44

L.H.S. : 112 is 12 more than the base (100)
We add the excess to the number
$112 + 12 = 124$
$124 + 1 (carry) = 125$

Answer: $112^2 = 12544$

Example 3 : 107^2

107^2

114 / 49

107 is 7 more than the base 100, so
<u>R.H.S</u>: $7^2 = 49$
<u>L.H.S.</u>: $107 + 07 = 114$

Answer: $107^2 = 11449$

Example 4 : 121^2

121^2

146 / 41

121 is 21 more than the base 100, so
<u>R.H.S</u>: $21^2 = 441$
<u>L.H.S</u> : $121 + 21 = 142$
$142 + 4 = 146$

Answer : $121^2 = 14641$

If we observe carefully we see that we can find the squares of number close to the base even orally. If the number is more than the base we add its complement to the given number and write the square of the complement next to it.

(Number + surplus) / Square of surplus

<u>Example:</u> $108^2 = (108 + 8) / 8^2$
$= 116 / 64$
$= 11664$

Given below are a few more examples solving sums in straight steps :

Example 5 : 109^2

$109^2 = 109 + 09 / 9^2$
$= 118 / 81$

Answer : **11881**

Example 6 : 1016^2

$1016^2 = 1016 + 016 / 16^2$
$= 1032 / 256$

Answer : **1032256**

Example 7 : 135^2

$135^2 = 135 + 35 / 35^2$
$= 170 / {}_{12}25$
$= 18225$

Answer : **18225**

Example 8 : 171^2

$171^2 = 171 + 71 / 71^2$
$= 242 / {}_{50}41$
$= 29241$

Answer : **29241**

When we compare both the methods we find that the Vedic method makes our calculations very short.
123^2

Usual method	Vedic method
123	$123 + 23 / 23^2$
x 123	$= 146 / 29$
----------	$= 15129$
369	
2460	
12300	

15129	

Square of Numbers less than the base :

For finding out squares of numbers less than the base we follow the same method, here instead of adding the surplus we subtract the deficiency.

Example 1: 97^2

Step 1:

/ 09

Base for 97 is 100 and it is 3 less than the base. We get the answer in 2 parts.

R.H.S. :We take the square of the complements
$$3^2 = 9$$
But 100 is the base so 2 digits on right side. So we write
9 as 09

Step 2:

94 / 09

L.H.S. : 97 is 3 less than the base (100)
We subtract the deficiency from the given number.
$$97 - 3 = 94$$

Answer : $97^2 = 9409$

Example 2 : 88^2

Step 1:

/ ₁44

Base for 88 is 100 and it is 12 less than the base. We get the answer in 2 parts

R.H.S. :We take the square of the complements
$$12^2 = {}_144$$
But 100 is the base so 2 digits on right side. So we write
44 and *carry over* 1 to right side

Step 2:

77 / 44

L.H.S. : 88 is 12 less than the base (100)
Subtract the deficiency from the given number.
$$88 - 12 = 76$$
$$76 + 1 \ (carry) = 77$$

Answer : $88^2 = 7744$

Example 3 : 9983^2

9983^2

$9976 / 289$

9983 is 17 less than the base.
R.H.S. : $17^2 = 289$
L.H.S. : $9983 - 17 = 9976$

Answer : $9983^2 = 9976289$

Example 4 : 82^2

$64 / {}_324$

$= 6724$

82 is 18 less than the base 100.
R.H.S. : $18^2 = {}_324$
L.H.S. : $82 - 18 = 64 + 3$ *(carry)* $= 67$

Answer : $82^2 = 6724$

After observing carefully we see if the number is less than the base we subtract its complements from the number and write the square of the complements next to it.

(Number – deficiency) / Square of deficiency

Example: $93^2 = (93 - 7)/7^2$
                    $= 8649$

Given below are a few more examples solving sums using straight steps:

Example 5 : 79^2

$79^2 = 79 - 21 / 21^2$
       $= 58 / 41$
       $= 6241$

Answer : **6241**

Example 6 : 86^2

$86^2 = 86 - 14 / 14^2$
      $= 72 / 96$
      $= 7396$

Answer : **7396**

Example 7 : 9989 ^{2}

9989 ^{2} = (9989 – 11) / 11^2
      = 9978 / 0121

Answer : **99780121**

Example 8 : 99885 ^{2}

99885 ^{2} = (99885 – 115) / 115^2
        = 99770 / 13225

Answer : **9977013225**

When we compare both the methods we find that the Vedic method makes our calculations very short.

987 ^{2}

Usual method

```
     987
   x 987
------------
    6909
   78960
  888300
------------
  974169
------------
```

Vedic method

987- 13 / 13 ^{2}
= 974169

Rules :

(1) Find the complements of the number from its nearest base.

(2) Right side of the answer is the square of the complements.

(3) If the number is less than the base, then we get the left side of the answer by subtracting the complements from the number.

(4) If the number is more than the base, then we get the left side of the answer by adding the complements to the number.

Exercise 12 .6 :
Try and solve these sums mentally :-

(1) 98^2

(2) 113^2

(3) 93^2

(4) 118^2

(5) 89^2

(6) 984^2

(7) 1007^2

(8) 993^2

(9) 9989^2

(10) 1022^2

Answers

Exercise 12 .1

(1) 625	(2) 4225	(3) 2025	(4) 7225
(5) 11025	(6) 42025	(7) 13225	(8) 50625
(9) 255025	(10)1010025		

Exercise 12 .2

(1) 4216	(2) 5621	(3) 7209	(4) 2024
(5) 13221	(6) 46224	(7) 99221	(8) 198009
(9)1243216	(10) 1010024		

Exercise 12 .3

(1) 441	(2) 2601	(3) 5041	(4) 8281
(5) 10201	(6) 12321	(7) 2801	(8) 63001
(9) 44521	(10) 1442401		

Exercise 12 .4

(1) 1936	(2) 4096	(3) 5476	(4) 8836
(5) 17956	(6) 23716	(7) 86436	(8)64516
(9) 295936	(10) 910116		

Exercise 12 .5

(1) 2116	(2) 4356	(3) 7396	(4) 9216
(5) 15876	(6) 30976	(7) 81796	(8) 106276
(9) 320356	(10) 913936		

Exercise 12 .6

(1) 9604	(2) 12796	(3) 8649	(4) 13924
(5) 7921	(6) 968256	(7) 1014049	(8) 986049
(9) 99780121	(10) 1044484		

Rectangular Box

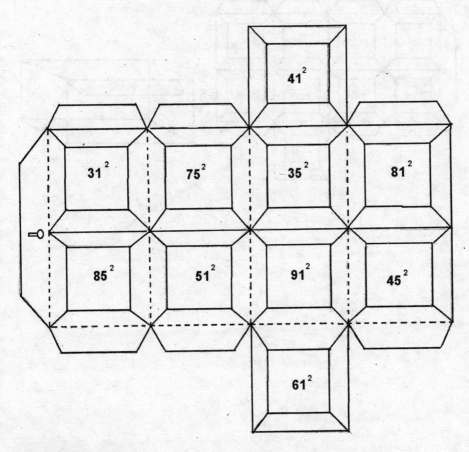

Find the squares.
If the answer is :
less than 3000
more than 3000

Colour the shape :
red
green

Fold it and make it into a box

Cut out the pattern along the outer lines and fold along
the dotted lines. Tape each flap to the under side of
either a rectangle or a square shape.

Answers to Rectangular Box

Final Shape

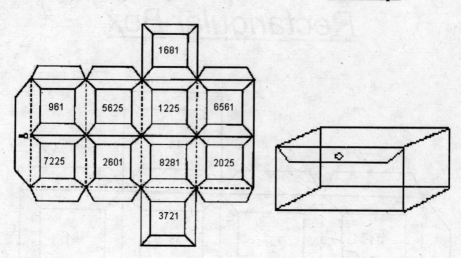

Rising Squares

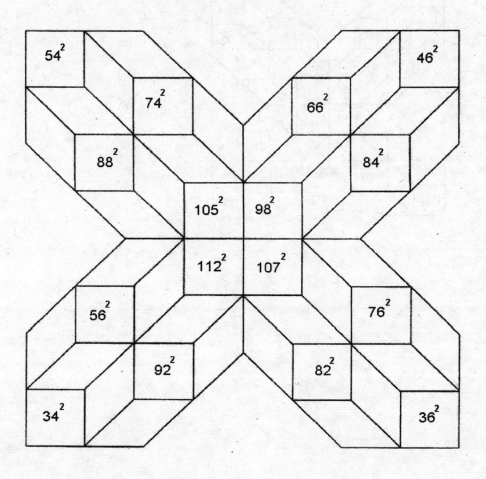

Find the squares.

If the answer is :

below 3000

between 3000-9000

more then 9000

colour the shape:

yellow

green

blue

Answers to Rising Squares

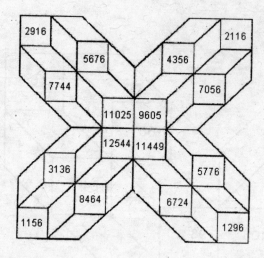

Chapter 13

STRAIGHT SQUARING

In the previous chapter we have studied the squares of specific numbers. In this chapter we will learn to find squares of any random number. To find out squares of numbers we will use the

द्वन्दव योग

"Dwandwa yoga"

"Duplex combination process"

To find our Duplex of numbers let us take a few examples:
We denote the Duplex by the symbol D.

(1) Duplex of a single digit number is its square.

$D(4) = 4^2 = 16$
$D(7) = 7^2 = 49$
$D(6) = 6^2 = 36$

(2) Duplex of a double-digit number is equal to twice the product of both the numbers as shown by the dot diagram.

$D(3\ 2) = 2(3 \times 2) = 12$
$D(3\ 6) = 2(3 \times 6) = 36$
$D(7\ 2) = 2(7 \times 2) = 28$
$D(4\ 8) = 2(4 \times 8) = 64$

(3) Duplex of a triple digit number is twice the product of the first and last digit plus the square of the middle digit as shown in the dot diagram. (For a triple digit number we make a pair of the first and the

last digit and take it as a double-digit number and the middle digit as
a single digit number and then find their duplex and add them.)

$D(135) = 2(1 \times 5) + 3^2 = 10 + 9 = 19$
$D(213) = 2(2 \times 3) + 1^2 = 12 + 1 = 13$
$D(347) = 2(3 \times 7) + 4^2 = 42 + 16 = 58$
$D(403) = 2(4 \times 3) + 0^2 = 24 + 0 = 24$

(4) Duplex of a four-digit number is twice the product of first and last
digit plus twice the product of second and third digit as shown in
the dot diagram.

$D(1234) = 2(1 \times 4) + 2(2 \times 3) = 20$
$D(4215) = 2(4 \times 5) + 2(2 \times 1) = 44$
$D(7362) = 2(7 \times 2) + 2(3 \times 6) = 64$
$D(3905) = 2(3 \times 5) + 2(9 \times 0) = 30$

(5) Duplex of a five digit number – Twice the product of the first and
last digit plus twice the product of the second and fourth digit plus
square of the third digit as shown in the dot diagram.

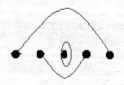

$D(12345) = 2(1 \times 5) + 2(2 \times 4) + 3^2 = 10 + 16 + 9 = 35$
$D(38154) = 2(3 \times 4) + 2(8 \times 5) + 1^2 = 24 + 80 + 1 = 105$

Rules :

    (1) Duplex of a single digit number is its square.

    (2) Duplex of a double digit number is twice the product of both
        the digits.

    (3) For bigger numbers we follow the rule of making pairs.

    (4) To make pairs we take the first digits on both the sides of the
        number.

    (5) We take twice the product of both the numbers in the pair.

    (6) If a single digit is in the middle then we take its square, if we
        get 2 pairs we again take twice the product of both the
        numbers in the pair.

    (7) We then add all the products.

**The dot diagrams below show the method to make pairs of bigger
numbers.**

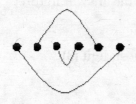
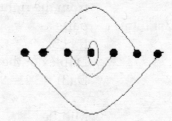

Exercise 13 .1 :

Find the Duplex of the following numbers :-

(1) 7   (6) 216

(2) 9   (7) 537

(3) 37   (8) 4126

(4) 72   (9) 7215

(5) 89   (10) 21532

Square of any Number using Duplex

To find the square of any number we use the *"Dwandwa yog"* along with our sutra

ऊर्ध्वतिर्यग्भ्याम्

"Ūrdhvatiryagbhyām"

"Vertical and Crosswise"

Here we will use only crosswise part of the sutra. In case of multiplication we start from the right column and keep adding a column in each step. When we have taken all the columns we start leaving one column at a time from the right again till we are left with the left most column. Here also we do the same with the digits of the number and keep taking the duplex of the digits as we move from right to left.
Let us understand it better with a few examples.

Example 1 : 23^2

Step 1:

23^2

\ D(last digit)

/ / 9

This is a double digit number so starting from the right we get the answer in three parts.

R.H.S. :
Duplex of the right most digit (3)
$D(3) = 3^2 = 9$

Step 2:

$2\ 3^2$

D(both digits)

/ $_1$2 / 9

Middle :
Duplex of both the digits of 23
$D(23) = 2\,(2 \times 3) = {}_1 2$

We write down 2 and *carry over* 1 to the next step.

Step 3:

23^2

/ D(first digit)

. / 9

L.H.S. :
Duplex of the left most digit 2
$D(2) = 2^2 = 4$
$4 + 1(carry) = 5$
Answer : $23^2 = 529$

Example 2 : 34^2

Step 1:

$3\ 4^2$

$\searrow$ D(last digit)

$/\ \ /\ {}_16$

This is a double digit number, so we get answer in three parts

R.H.S. :

Duplex of right most digit 4

$D(4) = 4^2 = {}_16$

We write down 6 and *carry over* 1 to next step

Step 2:

$3\ 4^2$

$\downarrow$ D(both digits)

$/\ {}_2 5\ /\ 6$

Middle :

Duplex of both the digits of 34

$D(34) = 2\ (3 \times 4) = 24$

$24 + 1\ (carry) = {}_25$

We write down 5 and *carry over* 2 to the next step

Step 3:

$3\ 4^2$

$\swarrow$ D(first digit)

$11\ /\ 5\ /\ 6$

L.H.S. :

Duplex of left most digit 3

$D(3) = 3^2 = 9$

$9 + 2\ (carry) = 11$

Answer : $34^2 = \mathbf{1156}$

Example 3 : 67^2

Step 1:

$6\ 7^2$

$\searrow$ D(last digit)

$/\ \ /\ {}_49$

R.H.S. :

$D(7) = 7^2 = {}_49$

We write down 9 and *carry over* 4 to next step

Step 2:

$6\ 7^2$

$\downarrow$ D(both digits)

$/\ {}_88\ /\ 9$

Middle

$D(67) = 2\ (6 \times 7) = 84$

$84 + 4\ (carry) = {}_88$

We write down 8 and *carry over* 8 to the next step

Step 3:

$6\ 7^2$

$\swarrow$ D(first digit)

$44\ /\ 8\ /\ 9$

L.H.S.

$D(6) = 6^2 = 36$

$36 + 8\ (carry) = 44$

Answer : $67^2 = \mathbf{4489}$

Example 4 : 73^2

     7 3^2

53 $_4$2 9

(1) $D(3) = 3^2 = 9$
(2) $D(73) = 2(7 \times 3) = {}_42$
(3) $D(7) = 7^2 = 49 + 4 = 53$

Answer : $73^2 = 5329$

Given below a few more examples solving sums in straight steps:

Example 5 : 54^2

$54^2 = D(5) / D(54) / D(4)$
$\quad = 5^2 / 2(5 \times 4) / 4^2$
$\quad = 25 / {}_40 / {}_16$
$\quad = 2916$

Answer : **2916**

Example 6 : 72^2

$72^2 = 7^2 / 2(7 \times 2) / 2^2$
$\quad = 49 / {}_28 / 4$
$\quad = 5184$

Answer : **5184**

Example 7 : 63^2

$63^2 = 36 / {}_36 / 9$
$\quad = 3969$

Answer : **3969**

Example 8 : 87^2

$87^2 = 64 / {}_{11}2 / {}_49$
$\quad = 7569$

Answer : **7569**

Exercise 13 .2 :

Find the squares of the following numbers :-

(1) 23^2

(2) 68^2

(3) 47^2

(4) 32^2

(5) 29^2

(6) 67^2

(7) 82^2

(8) 77^2

(9) 49^2

(10) 73^2

String Weaving

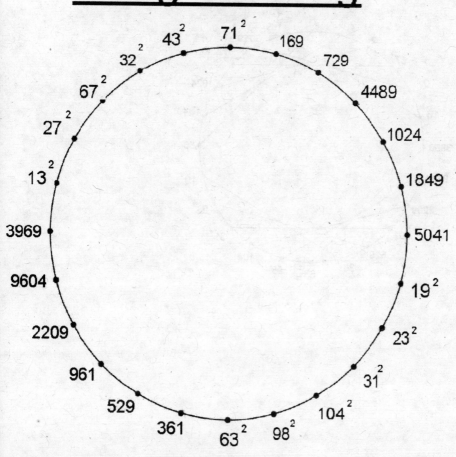

Find the squares and join their dots
to the dots of their answer.

Answers to String Weaving

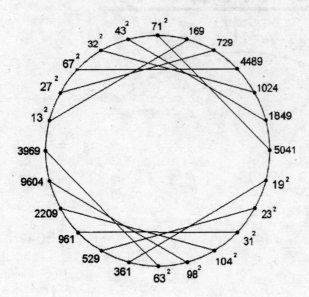

Squares of Triple digit Numbers

Example 9 : 152^2

       1 5 2

This is a triple digit number so we get the answer in 5 parts. Starting from the right.

Step 1:

152^2

     $\downarrow$ D(last digit)

/ / / / 4

Duplex of the last digit

$D(2) = 2^2 = 4$

Step 2:

$1\ 52^2$

   Y

   $\downarrow$ D(last 2 digits)

/ / /$_2$0 / 4

Duplex of two digits from the left

$D(52) = 2(5 \times 2) = {}_2 0$

Step 3:

152^2

    Y

    $\downarrow$ D(all 3 digits)

/ /$_3$1 / 0 / 4

Duplex of all the three digits

$D(152) = 2(1 \times 2) + 5^2 = 29$

$29 + 2\ (carry) = {}_3 1$

Step 4:

$15\ 2^2$

   Y

   $\downarrow$ D(first 2 digits)

/$_1$3 / 1 / 0 / 4

Duplex of two digits from the right

$D(15) = 2(1 \times 5) = 10$

$10 + 3\ (carry) = {}_1 3$

Step 5:

152^2

   / D(first digit)

   $\downarrow$

2 / 3 / 1 / 0 / 4

Duplex of the first (left most) digit

$D(1) = 1^2 = 1$

$1 + 1\ (carry) = 2$

Answer : $152^2 = \mathbf{23104}$

Example 10 : 273^2

$$273$$

This is a triple digit number so we get the answer in 5 parts. Starting from the right.

Step 1:

$$273^2$$

$\searpow$ D(last digit)

$/ / / / 9$

Duplex of the last digit

$D(3) = 3^2 = 9$

Step 2:

$$2\ 73^2$$

$\downarrow$ D(last 2 digits)

$/ / /_4 2 / 9$

Duplex of two digits from the left

$D(73) = 2(7 \times 3) = {}_4 2$

Step 3:

$$273^2$$

$\downarrow$ D(all 3 digits)

$/ /_6 5 / 2 / 9$

Duplex of all the three digits

$D(273) = 2(2 \times 3) + 7^2 = 61$

$61 + 4 \ (carry) = {}_6 5$

Step 4:

$$27\ 3^2$$

$\downarrow$ D(first 2 digits)

$/_3 4 / 5 / 2 / 9$

Duplex of two digits from the right

$D(27) = 2(2 \times 7) = 28$

$28 + 6 \ (carry) = {}_3 4$

Step 4:

$$273^2$$

$\downarrow$ D(first digit)

$7 / 4 / 5 / 2 / 9$

Duplex of the first digit

$D(2) = 2^2 = 4$

$4 + 3 \ (carry) = 7$

Answer : $273^2 = \mathbf{74529}$

We can solve a triple digit square in single step as below

Example 11 : 213^2

213^2

$4 / 4 / {}_13 / 6 / 9$

$= 4 / 5 / 3 / 6 / 9$

$= 45369$

(1) $D(3) = 3^2 = 9$
(2) $D(13) = 2 (1 \times 3) = 6$
(3) $D(213) = 2 (2 \times 3) + 1^2 = {}_13$
(4) $D(21) = 2 (2 \times 1) = 4$
$\quad 4 + 1 = 5$
(5) $D(2) = 2^2 = 4$

Answer : $213^2 = $ **45369**

Example 12 : 543^2

543^2

$= 25 / {}_40 / {}_46 / {}_24 / 9$

$= 29 / 4 / 8 / 4 / 9$

$= 294849$

(1) $D(3) = 3^2 = 9$
(2) $D(43) = 2 (4 \times 3) = {}_24$
(3) $D(543) = 2 (5 \times 3) + 4^2 = 46$
$\quad 46 + 2 = {}_48$
(4) $D(54) = 2 (5 \times 4) = 40$
$\quad 40 + 4 = {}_44$
(5) $D(5) = 5^2 = 25$
$\quad 25 + 4 = 29$

Answer : $543^2 = $ **294849**

Given below are a few more examples solving sums in straight steps :

Example 13 : 326^2

$= 9 / {}_12 / {}_40 / {}_24 / {}_36$

$= 106276$

Answer : **106276**

Example 14 : 427^2

$= 16 / {}_16 / {}_60 / {}_28 / {}_49$

$= 182329$

Answer : **182329**

Example 15 : 672^2

$= 36 / {}_84 / {}_73 / {}_28 / 4$

$= 451584$

Answer : **451584**

Example 16 : 867^2

$= 64 / {}_96 / {}_148 / {}_84 / {}_49$

$= 751689$

Answer : **751689**

Square of a Four digit Number

We follow the same pattern as shown below:

Example 17 :

2314^2 = D (2) / D (23) / D (231) / D (2314) / D (314) / D (14) / D (4)

= 2^2/2(2×3)/2(2×1) +3^2 /2(2×4) + 2(3×1)/2(3×4) +1^2/2(1×4) / 4^2

= 4 / ₁2 / ₁3 / ₂2 / ₂5 / 8 / ₁6

= 5354596

Example 18 :

2736^2 = D (2) / D (27) / D (273) / D (2736) / D (736) / D (36) / D (6)

= 2^2/2(2×7)/2(2×3) +7^2/2(2×6) + 2(7×3)/2(7×6) + 3^2/2(3×6)/6^2

= 4 / ₂8 / ₆1 / ₆6 / ₉3 / ₃6 / ₃6

= 7485696

Square of a five-digit number

Example 19 :

31052^2 = D(3) / D(31) / D(310) / D(3105) / D(31052) / D(1052) /

                    D(052) / D(52) / D(2)

= 3^2 /2(3x1)/2(3x0) + 1^2/ 2(3x5) + 2(1x0)/2(3x2) + 2(1x5) +0^2

= 9 / 6 / 1 / ₃0 / ₂2 / 4 / ₂5 / ₂0 / 4

= 964226704

Rules :

(1) We start writing our answer from the right side and move towards the left.

(2) We start with the duplex of the right most digit.

(3) We then keep adding a digit and finding their duplex.

(4) When we have taken all the digits then we start leaving digits from the right and go on till we are on the left most digit.

(5) While writing the answer we put down one digit at a time and carry over the other digits as carry over to the next step.

When we compare both the methods we find that the Vedic method makes our calculations very short.
473^2

Usual method

```
   473
 x 473
 ---------
  1419
 33110
189200
 ----------
223729
 ----------
```

Vedic method

$$= 16 \,/\, {}_5 6 \,/\, {}_7 3 \,/\, {}_4 2 \,/\, 9$$
$$= 223729$$

Exercise 13 .3 :

Find the squares of the following :-

(1) 263^2

(2) 385^2

(3) 237^2

(4) 723^2

(5) 876^2

(6) 486^2

(7) 763^2

(8) 561^2

(9) 203^2

(10) 6134^2

Answers

Exercise 13 .1			
(1) 49	(2) 81	(3) 42	(4) 28
(5) 144	(6) 25	(7) 79	(8) 52
(9) 74	(10) 39		

Exercise 13 .2			
(1) 529	(2) 4624	(3) 2209	(4) 1024
(5) 841	(6) 4489	(7) 6724	(8) 5929
(9) 2401	(10) 5329		

Exercise 13 .3			
(1) 69169	(2) 148225	(3) 56169	(4) 522729
(5) 767376	(6) 236196	(7) 582169	(8) 314721
(9) 41209	(10) 37625956		

Flexagon

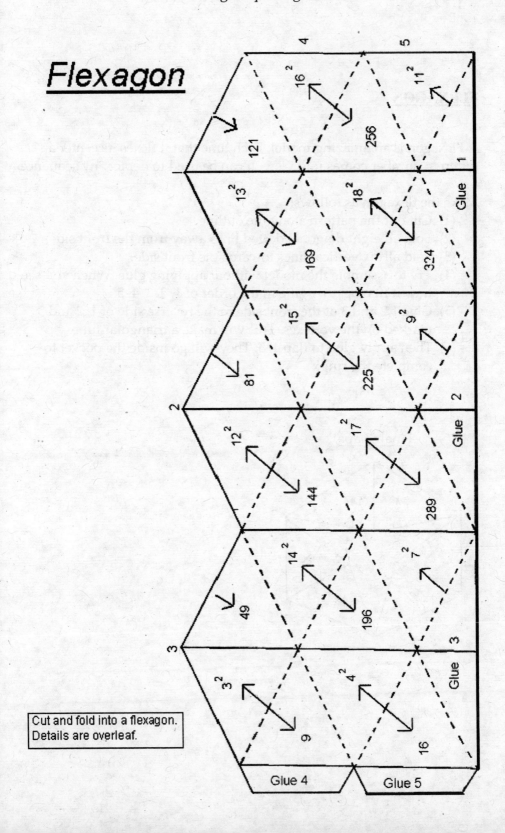

Cut and fold into a flexagon.
Details are overleaf.

FLEXAGON

A Flexagon is an amazing model. Each time that it flexes its center a different number comes into view. It can be used to depict any sequence.

Make the flexagon as follows:
  (1) Cut out the pattern along the outline.
  (2) Fold all eight diagonal dotted lines away from the front side.
  (3) Fold all six vertical lines towards the front side.
  (4) Try to assemble the model without applying glue. When you see how it fits together glue in the order of 1, 2, 3, 4, 5.
  (5) Glue 1,2 and 3 of the front side to the reverse side of 1, 2 and 3 marked on the vertexes. This will make a triangular tube.
  (6) Then apply glue to flap 4, 5. They will go inside the pocket to complete the ring.

Final Shape

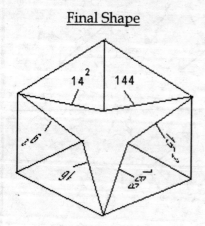

Chapter 14

CUBES

After learning squares of number, now we will learn to find the cubes of two-digit numbers. For this we use the sutra

श्रानुरूप्येण

"Ānurūpyeṇa"

"Proportionally"

This sutra makes the cubing very easy. All we need to know is the cubes of the first 10 natural numbers and we can find the cube of any two digit number. We first write the cube of the first digit of the number and then write the next three numbers in geometric ratio in the exact proportion as the digits of the number. Below the second and the third numbers we write their double respectively. Then we just add and keep writing one digit per number and carry over the other digits to the next number.

Let us understand it further with the help of a few examples.

__Example 1__ : 12^3

__Step 1:__	The two digits of the number 12 are in the ratio of
1 2 4 8	1 : 2
	Now we write the cube of 1, $1^3 = 1$
	The next three numbers in the same ratio of
	1: 2 for 1 are 2, 4, 8

__Step 2:__	Double of 2nd and 3rd term are
	$2 \times 2 = 4$
1 2 4 8	$2 \times 4 = 8$
4 8	We write them below 2 and 4 respectively.

<u>Step 3:</u>

```
1  2  4  8
      4  8
---------------
1  7 ₁2  8
---------------
```

We add the terms and write one digit at a time

(1) $8 = 8$
(2) $4 + 8 = {}_12$
(3) $2 + 4 = 6$
     $6 + 1 \, (carry) = \, 7$
(4) $1 = 1$

Answer : $12^3 = 1728$

<u>**Example 2** : 14^3</u>

<u>Step 1 :</u>

```
1  4  16  64
```

The two digits of the number are in the ratio of $1 : 4$

Now we write the cube of 1, $1^3 = 1$
The next three numbers in the same ratio of
$1 : 4$ for 1 are 4, 16, 64

<u>Step 2:</u>

```
1  4  16  64
   8  32
```

Double of 2nd and 3rd term are

$2 \times 4 = 8$
$2 \times 16 = 32$
We write them below 4 and 8

<u>Step 3:</u>

```
1  4  16  64
   8  32
--------------------
2 ₁7 ₅4 ₆4
--------------------
```

After addition we write down the numbers by putting down one digit at a time

(1) $64 = {}_64$
(2) $16 + 32 = 48$
     $48 + 6 \, (carry) = {}_54$
(3) $4 + 8 = 12$
     $12 + 5 \, (carry) = {}_17$
(4) $1 + 1 \, (carry) = \, 2$

Answer : $14^3 = 2744$

From the above two examples we conclude some interesting facts and this makes our cubing more easy.

(1) In the first row we are writing the four numbers in geometric ratio in exact proportion. This means that if a, b are the two digits of our number then the four terms are

$$a^3 \qquad a^2 \times b \qquad a \times b^2 \qquad b^3$$

(2) In the next step we write the double of the middle two terms.

$$a^3 \qquad a^2 \times b \qquad a \times b^2 \qquad b^3$$
$$2(a^2 \times b) \qquad 2(a \times b^2)$$

(3) We now add the two middle terms and copy down the first and last term as it is.

$$a^3 \qquad a^2 \times b \qquad a \times b^2 \qquad b^3$$
$$2(a^2 \times b) \qquad 2(a \times b^2)$$

$$a^3 \qquad 3(a^2 \times b) \qquad 3(a \times b^2) \qquad b^3$$

We see that this is nothing but the breaking up of our formula for $(a + b)^3$ where a, b are two individual digits of the number. The simplification and breaking up of the formula makes our calculations very fast and easy.

Let us take a few more examples and understand its simplicity.

Example 3 : 23^3

Here we observe that the ratio between the digits is 2 : 3 and calculating the next three digits will be cumbersome, so we follow the new method.

Step 1:   Here a = 2 and b = 3,
                                The first four terms will be
                                $a^3 = 8$
8 12 18 27 $a^2 b = 2^2 \times 3 = 12$
                                $a b^2 = 2 \times 3^2 = 18$
                                $b^3 = 27$

Step 2:

8 12 18 27
    24 36

We now double the 2nd and the 3rd terms.
$2 \times 12 = 24$
$2 \times 18 = 36$
We write them below the 2nd and the 3rd terms.

Step 3:

8 12 18 27
   24 36
----
12 $_4$1 $_5$6 $_2$7

Add them and place one digit at a time.
(1) $27 = {}_2 7$
(2) $18 + 36 = 54$
    $54 + 2$ (carry) $= {}_5 6$
(3) $12 + 24 = 36$
    $36 \quad 5$ (carry) $= {}_4 1$
(4) $8 + 4$ (carry) $= 12$

Answer : $23^3 = \mathbf{12167}$

Example 4 : 17^3

Step 1:

1 7 49 343

Here a = 1 and b = 7

The first four terms will be
$a^3 = 1$
$a^2 b = 1^2 \times 7 = 7$
$a b^2 = 1 \times 7^2 = 49$
$b^3 = 343$

Step 2:

1 7 49 343
  14 98

We now double the 2nd and the 3rd terms.
$2 \times 7 = 14$
$2 \times 49 = 98$
We write them below the 2nd and the 3rd terms.

Step 3:

1 7 49 343
  14 98
----
4 $_3$9 $_{18}$1 $_{34}$3

Add them and place one digit at a time.
(1) $343 = {}_{34} 3$
(2) $49 + 98 = 147$
    $147 + 34$ *(carry)* $= {}_{18} 1$
(3) $7 + 14 = 21$
    $21 + 18$ *(carry)* $= {}_3 9$
(4) $1 + 3$ *(carry)* $= 4$

Answer : $17^3 = \mathbf{4913}$

Example 5 : 25^3

8	20	50	125
	40	100	

$15 \ _7 6 \ _{16} 2 \ _{12} 5$

The four terms will be 8, 20, 50, 125

Double of middle two terms is 40, 100

Adding all terms and placing one digit at a time.

(1) $125 = {}_{12}5$

(2) $50 + 100 = 150$

$\qquad 150 + 12 (carry) = {}_{16}2$

(3) $20 + 40 = 60$

$\qquad 60 + 16(carry) = {}_7 6$

(4) $8 + 7(carry) = 15$

Answer : $25^3 = \mathbf{15625}$

Rules :

(1) The method of finding cubes using the geometric ratio in exact proportion is easy if the ratio between both the digits of the number is simple but for a complicated ratio we use the breakup method.

(2) For the breakup method, we have to write four terms in the first row, if a, b are the two terms of the number,

   a) First term is the cube of the first digit (a^3).

   b) Second is the multiplication of the square of the first and the second digit, ($a^2 b$).

   c) Third is the multiplication of the first digit and the square of second digits ($a b^2$).

   d) Fourth is the cube of the second digit (b^3).

(3) We double the middle two terms and write them below the second and third terms respectively.

(4) We start putting down the answer by adding the middle two terms and keeping the first and last term as they are.

(5) We put down one digit at a time and carry over the other digits in each term to the previous term.

Given below are a few more examples solving sums in straight steps :

Example 6 : 43^3

$$
\begin{array}{cccc}
64 & 48 & 36 & 27 \\
 & 96 & 72 & \\
\hline
79\ _{15}5 & _{11}0 & _2 7 & \\
\hline
\end{array}
$$

Answer : 79507

Example 7 : 29^3

$$
\begin{array}{cccc}
8 & 36 & 162 & 729 \\
 & 72 & 324 & \\
\hline
24\ _{16}3 & _{55}8 & _{72}9 & \\
\hline
\end{array}
$$

Answer : 24389

Example 8 : 64^3

$$
\begin{array}{cccc}
216 & 144 & 96 & 64 \\
 & 288 & 192 & \\
\hline
262 & _{46}1 & _{29}4 & _6 4 \\
\hline
\end{array}
$$

Answer : **262144**

Example 9 : 32^3

$$
\begin{array}{cccc}
27 & 18 & 12 & 8 \\
 & 36 & 24 & \\
\hline
32 & _5 7 & _3 6 & 8 \\
\hline
\end{array}
$$

Answer : **32768**

Exercise 14 . 1 :

Find the cubes of the following numbers :-

(1) 22^3

(2) 31^3

(3) 35^3

(4) 27^3

(5) 41^3

(6) 51^3

(7) 62^3

(8) 54^3

(9) 72^3

(10) 37^3

Emerging Diamond

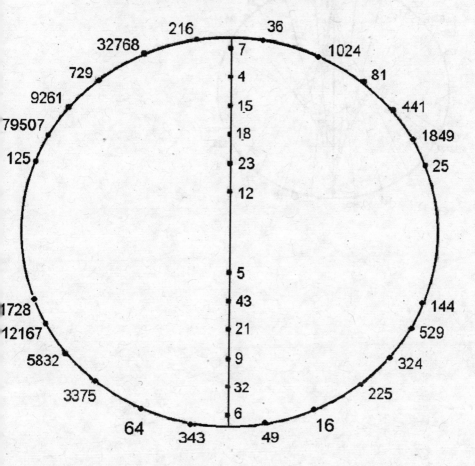

Join the numbers on the diameter of the circle with
their equivalent squares and cubes.

Answers to Emerging Diamond

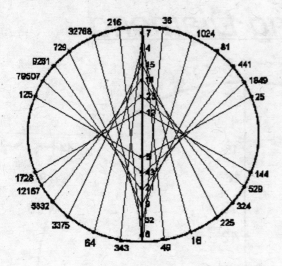

Flexagon

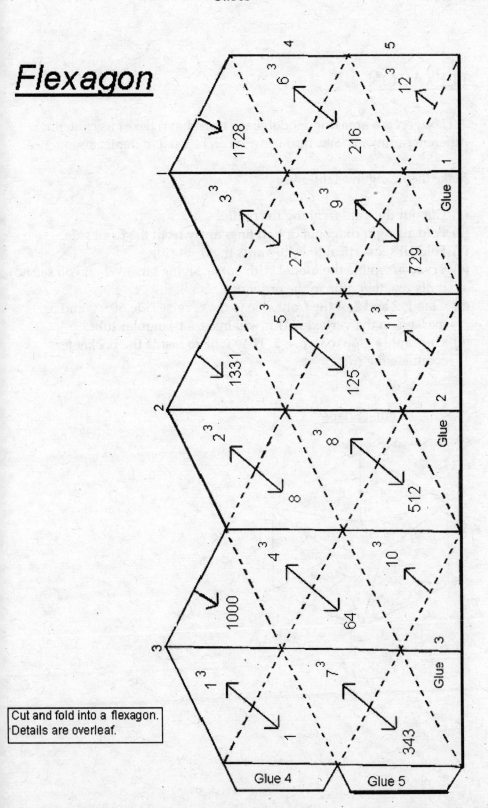

Cut and fold into a flexagon.
Details are overleaf.

FLEXAGON

A Flexagon is a amazing model. Each time that it flexes its center a different number comes into view. It can be used to depict a sequence.

Make the flexagon as follows :

(1) Cut out the pattern along the outline.
(2) Fold all eight diagonal dotted lines away from the front side.
(3) Fold all six vertical lines towards the front side.
(4) Try to assemble the model with out applying glue. When you see how it fits together glue in the order of 1, 2, 3, 4, 5.
(5) Glue 1, 2 and 3 of the front side to the reverse side of 1, 2 and 3 marked on the vertexes. This will make a triangular tube.
(6) Then apply glue to flap 4, 5. They will go inside the pocket to complete the ring.

Final Shape

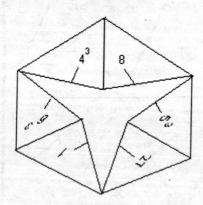

Cubes of Numbers close to the Base

In the earlier chapters also we have dealt with mathematical operations on different kinds of numbers, those close to the base and random numbers. For cubes also we have a different method of calculating cubes for those numbers close to the base. When our numbers are close to the bases, we can use our sutra

यावदूनं

"Yāvadūnam"

"By the deficiency"

First we will deal with **numbers more than the base**. Let us take a few examples to understand this.

Example 1 : 104 ^{3}

104 ^{3}

104 is 4 more than the base 100
We get our answer in 3 parts and this time we start from the left.

Step 1:

112 /   /

L.H.S. :
Number plus twice the excess
(104 is 4 more than the base 100)
$$104 + (2 \times 4)$$
$$= 104 + 8 = 112$$

Step 2:

112 / 48 /

Middle:
We multiply the excess of the L.H.S. of the answer with the excess of the number (112 is 12 more than the base).
$$12 \times 4 = 48$$
(100 is the base of the number so 2 digits in this part)

Step 3:

112 / 48 / 64

R.H.S. :
Cube of the excess of the number
$\quad 4^3 = 64$
(100 is the base of the number so 2 digits in this part)

Answer : $104^3 = 1124864$

Example 2 : 105^3

105^3

105 is 5 more than the base 100
We get our answer in 3 parts and we start from the left.

Step 1:

115 /   /

L.H.S. :
Number plus twice the excess
(105 is 5 more than the base 100)
$\quad 105 + (2 \times 5)$
$= 105 + 10 = 115$

Step 2:

115 / 75/

Middle:
We multiply the excess of the L.H.S. by the excess of the number (115 is 15 more than the base).
$\quad 15 \times 5 = 75$

(100 is the base of the number so 2 digits in this part)

Step 3 :

115 / 75 / $_1$25
----
115 / 76 / 25

R.H.S. :
Cube of the excess of the number
$\quad 5^3 = {}_1 25$

(100 is the base of the number so 2 digits in this part. We write 25 and *carry over* 1 to middle part).

Now Middle part : $75 + {}_1 = 76$

Answer : $105^3 = 1157625$

Example 3 : 1007^3

1007^3	1007 is 7 more than the base 1000 We get our answer in 3 parts and we start from the left.

Step 1:

1021 /   /

L.H.S. :
Number plus twice the excess
(1007 is 7 more than the base 1000)
  $1007 + (2 × 7)$
$= 1007 + 14 = 1021$

Step 2:

1021 / 147 /

Middle :
We multiply the excess of the LHS by the excess of the number (1021 is 21 more than the base).
  $21 × 7 = 147$

(1000 is the base of the number so 3 digits in this part)

Step 3:

1021 /147/ 343

R.H.S. :
Cube of the excess of the number
  $7^3 = 343$

(1000 is the base of the number so 3 digits in this part.)

Answer : $1007^3 = 1021147343$

Example 4 : 10012^3

10012^3	10012 is 12 more than the base 10000 We get our answer in 3 parts and we start from the left.

Step 1:

10036 /   /

L.H.S. :
Number plus twice the excess
  (10012 is 12 more than the base 10000)
    $10012 + (2 × 12)$
  $= 10012 + 24 = 10036$

Step 2:   Middle :
                               We multiply the excess of the LHS by the
10036 / 0432 /   excess of the number (10036 is 36 more than
                               the base)
                                   36 × 12 = 432

                               (10000 is the base of the number so 4 digits in
                               this part, as we have three we add a zero
                               before our number)
                                   432 = 0432

Step 3:   R.H.S. :
                               Cube of the excess of the number
10036 / 0432 / 1728   12 ^{3} = 1728

                               (10000 is the base of the number so 4 digits in
                               this part)

                               Answer : 10012 ^{3} = 1003604321728

Example 5 :121 ^{3}

                               L.H.S. :   121 + 42 = 163

163 / $_{13}$23 / $_{92}$61   Middle : 63 x 21 = $_{13}$23

= 1771561   R.H.S. :   21 ^{3} = 8 4 2 1
                                                   8 4
                                                ----
                                                   9 2 6 1

                               Now middle becomes 1323 + 92 = $_{14}$15

                               L.H.S. :   163 + 14 = 177

                               Answer : 121 ^{3} = 1771561

Till now we have discussed cubes of numbers more than the base, now let us take a few examples when our *numbers are less than the base*.

Example 6 : 98 ^{3}

98 ^{3}	98 is 2 less than the base 100 We get our answer in 3 parts and this time also we start from the left.

Step 1

94 /   /

L.H.S. :
Number plus twice the deficient
(98 is 2 less than the base 100)
    98 + (2 × - 2)
= 98 - 4
= 94

Step 2:

94 / 12/

Middle :
We multiply the deficient of the L.H.S. (94 is 6 less than the base) by the deficient of the number   - 6 × - 2 = 12

(100 is the base of the number so 2 digits in this part)

Step 3:

94 / 12 / $\overline{08}$

R.H.S. :
Cube of the deficiency of the number
    (- 2) ^{3} = - 8

(100 is the base of the number so 2 digits in this part.
So we write – 8 as $\overline{08}$

Step 4:

94 / 12 / $\overline{08}$
----
94 / 11/ 92

Now since we have a negative number in the R.H.S. we take its complement to make it positive and subtract 1 extra from the middle part for coming out of complements.

complement of 08 = 92
Middle part 12 - 1 = 11

Answer : 98 ^{3} = 941192

Example 7 : 94^3

94^3

94 is 6 less than the base 100
We get our answer in 3 parts and we start from the left.

Step 1:

82 /　　/

L.H.S. :
Number plus twice the deficient
(94 is 6 less than the base 100)
　$94 + (2 \times -6)$
$= 94 - 12$
$= 82$

Step 2:

82 / $_1$08 /

Middle :
We multiply the deficient of the L.H.S. by the deficient of the number (82 is 18 less than the base)
　$-18 \times -6 = {}_1 08$

(100 is the base of the number so 2 digits in this part)

Step 3 :

82 / $_1$08 / $_2\overline{16}$

R.H.S. :
Cube of the deficiency of the number

　　$(-6)^3 = -216 = {}_2\overline{16}$

(100 is the base of the number so 2 digits in this part. So we write -16 and *carry over* -2 (negative carry over) to the middle part.

82 / $_1$06 / $\overline{16}$

So middle part becomes
　$108 - 2 \,(carry) = {}_106$

Step 4:

82 / $_1$06 / $\overline{16}$

----

83 / 05 / 84

Now since we have a negative number in the R.H.S. we take its complement to make it positive and subtract 1 extra from the middle part for coming out of complements.

complement of 16 = 84
Middle : 106 – 1 (for complement) = $_105$
L.H.S. : 　82 + 1 = 83

Finally,
    R.H.S. : 84
    Middle : 05
    L.H.S. : 83

Answer : 94^3 = 830584

Example 8 : 9991^3

9991^3

9991 is 9 less than the base 10000
We get our answer in 3 parts and we start from the left.

Step 1:

9973 / /

L.H.S. :
Number plus twice the deficient
(9991 is 9 less than the base 10000)
  $9991 + (2 \times -9)$
= 9973

Step 2:

9973 / 0243 /

Middle :
We multiply the deficient of the L.H.S. (-27)
by the deficient of the number (- 9)
  $-27 \times -9 = 0243$
(10000 is the base of the number so 4 digits in this part.)

Step 3:

9973 / 0243 / $\overline{0729}$

R.H.S. :
Cube of the deficiency of the number

  $(-9)^3 = -0729 = \overline{0729}$

(10000 is the base of the number so 4 digits in this part.
Now since we have a negative number in the R.H.S we take its complement to make it positive and subtract 1 extra from the middle part for coming out of complements

Step 4:

9973/ 0243 /$\overline{0729}$

9973 0242 9271

complement of 0729 = 9271
Middle 0243 – 1 (for complement) = 0242

Answer : 9991^3 = **997302429271**

Example 9 : 83 ^{3}

83 ^{3}

———

49 / $_8$67 / $_{49}$13

57 / 17 / 87

(1) L.H.S. :   83 - 34 = 49

(2) Middle :   -51 x -17 = 867

(3) R.H.S. : $(-17)^3$ = 1 7 49 343

                                    14 98

----

                              4 $_3$9 $_{18}$1 $_{34}$3

                              ———

                    = - 4913 = $_{49}$13

complement of 13 = 87

Finally

R.H.S. : 87

Middle :   867 –49 (negative carry) = 818

                818 – 1 (for complements) = $_8$17

L.H.S. :   49 +8 = 57

Answer : 83 ^{3} = 571787

Rules :

(1) When a number is close to a base we find its excess or deficiency from its nearest base.

(2) We get our answer in three parts starting from the left.

(3) Left part is the addition of twice the excess/deficiency to the number itself.

(4) Middle part is the multiplication of

- The new excess (of the answer of the left part from the base) with the old excess (of the given number from the base) when the number is more than the base.

- The new deficiency (of the answer of the left part from the base) with the old deficiency (of the given number from the base) when the number is less than the base.

(5) Right part is the cube of the excess (when the given number is more than the base) OR cube of the deficiency (when the number is less than the base.

(6) The base of the given number determines the number of digits in the middle and R.H.S. As many zeros in the base, those many digits in these two parts.

Given below are a few more examples solving sums in straight steps :

Example 10 : 113^3

$$= (113 + 26)/\ (13 \times 39)/\ 13^3$$
$$= 139\ /\ _5\ 07\ /\ _{21}\ 97$$
$$= 1442897$$

Answer : **1442897**

Example 11 : 132^3

$$= (132 + 64)/\ (96 \times 32)\ /\ 32^3$$
$$= 196\ /\ _{30}\ 72\ /\ _{327}\ 68$$
$$= 2299968$$

Answer : **2299968**

Example 12 : 82^3

$$= (82-36)\ /\ (-54 \times -18)\ /\ (-18)^3$$

$$= 46\ /\ _9\ 72/\ _{58}\overline{32}$$

$$= 55\ /\ 14\ /\overline{32}$$
$$= 551368$$

Answer : **551368**

Example 13 : 99992^3

$$= (99992-16)/(-8 \times -24)/\ (-8)^3$$

$$= 99976\ /\ 00192\ /\ \overline{00512}$$
$$= 99976\ /\ 00191\ /\ 99488$$
$$= 999760019199488$$

Answer : **999760019199488**

When we compare both the methods we find that the Vedic method makes our calculations very short.

39^3

Usual method		Vedic method

Usual method
```
     39
   x 39
  ---------
    351
   1170
  ---------
   1521
   x 39
  -----------
  13689
  45630
  -----------
  59319
```

Vedic method
```
27  81  243  729
    162  486
------------------------
59 ₃₂3 ₈₀1   ₇₂9
------------------------
      59319
```

If the number is 106 the usual method is the same, but Vedic is much simpler

$106^3 = 106 + 12 / 18 \times 6 / 6^3$

$= 119 / {}_1 10 / {}_2 16$

$= 1191016$

Exercise 14 . 2 :

Find the cubes of numbers close to their bases :-

(1) 105^3 　　　　　　(6) 9998^3

(2) 1002^3 　　　　　　(7) 10002^3

(3) 107^3 　　　　　　(8) 99993^3

(4) 99^3 　　　　　　(9) 996^3

(5) 97^3 　　　　　　(10) 100007^3

Answers

Exercise 14 .1

(1) 10648	(2) 29791	(3) 42875	(4) 19683
(5) 68921	(6) 132651	(7) 238328	(8) 157464
(9) 373248	(10) 50653		

Exercise 14 .2

(1) 1157625	(2) 1006012008	(3) 1225043	(4) 970299
(5) 912673	(6) 99940019992	(7) 1000600120008	
(8) 999790014699657	(9) 988047936	(10)1000210014700343	

Chapter 15

SQUARE ROOTS OF EXACT SQUARES

To find out the square root of exact squares first of all we should know if the number is a perfect square or not. To find out if any given number is an exact square we should follow the following fundamental rules:

(1) A perfect square ends in 0, 1, 4, 5, 6 & 9.
(2) A number is not a perfect square if it ends in 2, 3, 7 or 8.
(3) The number should end in even number of zeros.
(4) If the number ends in 6, then its second last digit should be odd.
(5) If the number does not end in 6 then its second last digit should be even.
(6) If the number is even its last two digits should be divisible by 4.
(7) If the given number has n digits then its square root will have, $n/2$ digits if n is even, or $(n+1)/2$ digits if n is odd.

Now let us study the squares of the first 10 natural numbers

Number	Square	Last digit of square
1	1	1
2	4	4
3	9	9
4	16	6
5	25	5
6	37	6
7	49	9
8	64	4
9	81	1
10	100	00

From the above table we conclude that :

(1) If the square ends in 1 then the last digit of its square root will either be 1 or 9.
(2) If the square ends in 4 then the last digit of its square root will either be 2 or 8.
(3) If the square ends in 5 then the last digit of its square root will either be 5 or 00.
(4) If the square ends in 6 then the last digit of its square root will either be 4 or 6.
(5) If the square ends in 9 then the last digit of its square root will either be 3 or 7.

After observing the above mentioned facts regarding the square roots finding out square root of exact squares becomes very easy now. We make groups of 2 digits in the given number from the right. The number of pairs we get, those many numbers of digits will be there in the square root. We use the sutra

विलोकनं

"Vilokanam"

"By mere observation"

Following the sutra by mere observing the above mentioned table we can find out the **Last digit** of our square root and we can find out the **First digit** of the square root by observing the first group of the number.

Let us understand it with a few examples :-

Example 1 : $\sqrt{1024}$

Step 1:

$\sqrt{10, 24}$

We make groups of 2 digits starting from the right. Here we have 2 groups.
So two digits in the square roots.

Step 2:

$\sqrt{10, 24}$

= __2 or __8

As the square ends in 4 ,
square root ends in 2 or 8 (by observing the table of squares).

So we just have two options in the unit place of our answer.

Step 3:

$\sqrt{10, 24}$

= 3 2 or 3 8

From the left group i.e. 10,
and since $3^2 = 9$ and $4^2 = 16$

Our number 10 lies between the two numbers, 9 and 16, so the first digit of our square root will be 3.

We cannot take 4 as $40^2 = 1600$, which is bigger than the given number 1024, so we take 3 as the first digit of our answer.

Step 4:

$\sqrt{10, 24} = 32$

We have reduced our options for the answer to 32 and 38.
Now we know that $35^2 = 1225$
(using "*Ekadhikena Purven*")

Since 1024 is an exact square and is less than 1225, so the square root will be less than 35
So the answer is 32.

Answer : $\sqrt{1024} = 32$

Example 2 : $\sqrt{2116}$

Step 1:

$\sqrt{21, 16}$

We make groups of 2 digits starting from the right. Here we have 2 groups.

Step 2:

$\sqrt{21, 16}$

= __4   or __6

As the square ends in 6 ,
square root ends in 4 or 6

So we just have two options in the unit place of our answer.

Step 3:

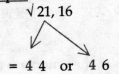

$\sqrt{21, 16}$

$= 4\ 4$ or $4\ 6$

Observe the left group number 21,

and since $4^2 = 16$ and $5^2 = 25$
Our number 21 lies between the two numbers 16 and 25.
So the first digit of our square root will be 4.

We cannot take 5 as $50^2 = 2500$, which is bigger than the given number 2116 so we take 4 as the first digit of our answer.

Step 4:

$\sqrt{2116} = 46$

We have reduced our options for the answer to 44 and 46
Now we know that $45^2 = 2025$
Since 2116 > 2025 and since it is a perfect square, the square root will be more than 45

So the square root is 46

Answer : $\sqrt{2116} = 46$

Example 3 : $\sqrt{17689}$

Step 1:

$\sqrt{176,89}$

We make groups of 2 digits starting from the right. Here we have 3 groups. But for the purpose of our calculation we take the first and the second group as one so we have 176 and 89 as the two groups.

Step 2:

$\sqrt{176,8\ 9} = \_\_3$
       or $\_\_7$

As the square ends in 9 ,
Square root ends in 3 or 7

Step 3:

$\sqrt{176\ ,\ 89}$

$= 13\ 3$ or $13\ 7$

Observe the left group number 176,

and since $13^2 = 169$ and $14^2 = 196$
Our number 176 lies between the two numbers 169 and 196.

So the first digits of our square root will be 13.

Step 4:

$\sqrt{176,89} = 133$

We have reduced our options for the answer to
133 and 137

Now we know that $135^2 = 18225$
Since $17689 < 18225$
so the square root will be less than 135
So the answer is 133

Answer : $\sqrt{17689} = 133$

Given below are a few more examples solving sums in straight steps :

Example 4 : $\sqrt{9,61}$

$\sqrt{9,61} = \underline{}1 / \underline{}9$
     $(3^2 = 9, 4^2 = 16)$
  $= 31 / 39$
     $(35^2 = 1225 \ \& \ 961 < 225)$
  $= 31$

Answer : 31

Example 5 : $\sqrt{27,04}$

$\sqrt{27,04} = \underline{}2 / \underline{}8$
     $(5^2 = 25, 6^2 = 36)$
  $= 52 / 58$
     $(55^2 = 3025 \ \& \ 2704 < 3025)$
  $= 52$

Answer : 52

Example 6 : $\sqrt{73,96}$

$\sqrt{73,96} = \underline{}4 / \underline{}6$
     $(8^2 = 64, 9^2 = 81)$
  $= 84 / 86$
     $(85^2 = 7225 \ \& \ 7396 > 7225)$
  $= 86$

Answer : 86

Example 7 : $\sqrt{44,89}$

$\sqrt{44,89} = \underline{}3 / \underline{}7$
     $(6^2 = 36, 7^2 = 49)$
  $= 63 / 67$
     $(65^2 = 4225 \ \& \ 4489 > 4225)$
  $= 67$

Answer : 67

Rules :

(1) Make groups of two digits in the number starting from the right.

(2) We find out the two options for the last digit in the answer by observing the last digit in the number.

(3) By observing the first pair of the number, we find out the first digit of the answer.

(4) We find out the square of the number ending in 5 (using our sutra "*Ekadhikena Purvena*") between our two options of the answer.

(5) Our number will be bigger or smaller than the square of the number ending in 5, accordingly we can choose the correct answer from the two options.

But this method is possible only if the number given to us is a perfect square and it is a four-digit number. We can use this method for five digits. If the square has more then five digits then this method is not very suitable. We will study square roots of larger numbers in the next chapter.

Exercise 15 :

Find out the square roots of the following perfect squares by mere observation :-

(1) $\sqrt{529}$

(2) $\sqrt{1089}$

(3) $\sqrt{2401}$

(4) $\sqrt{6724}$

(5) $\sqrt{3969}$

(6) $\sqrt{6241}$

(7) $\sqrt{23716}$

(8) $\sqrt{12544}$

(9) $\sqrt{17161}$

(10) $\sqrt{11881}$

Answers

Exercise 15			
(1) 23	(2) 33	(3) 49	(4) 82
(5) 63	(6) 79	(7) 154	(8) 112
(9) 131	(10) 109		

Chapter 16

CUBE ROOTS OF EXACT CUBES

Cube Roots of exact cubes can be found out by following the same pattern as that of the square roots. Let us first study the cubes of the first 9 natural numbers to determine the last digit of the cubes.

Number	Cubes	Last digit in the Cube
1	1	1
2	8	8
3	27	7
4	64	4
5	125	5
6	216	6
7	343	3
8	512	2
9	729	9

From the above table we conclude that :

(1) If the cube ends in 1 then the cube root (CR) ends in 1.
(2) If the cube ends in 2 then the cube root (CR) ends in 8.
(3) If the cube ends in 3 then the cube root (CR) ends in 7.
(4) If the cube ends in 4 then the cube root (CR) ends in 4.
(5) If the cube ends in 5 then the cube root (CR) ends in 5.
(6) If the cube ends in 6 then the cube root (CR) ends in 6.
(7) If the cube ends in 7 then the cube root (CR) ends in 3.
(8) If the cube ends in 8 then the cube root (CR) ends in 2.
(9) If the cube ends in 9 then the cube root (CR) ends in 9.

We observe that there is no overlapping of digits in the cubes of the first ten natural numbers as was in the case of square.

To find the cube root of a number we make groups of three digits starting from the right. The number of groups we get those many numbers of digits in the CR, irrespective of the number of digits in the left most group. We use the sutra

विलोकनं

"Vilokanam"

"By mere observation"

Following the sutra by mere observing the last digit of the cube we can find out the **Last digit** of the cube root. We can find the **First digit** of the cube root by mere observation of the first group in the number. This method can be used to find the cube roots of any number which has up to six digits. Let us understand it with a few examples.

Example 1 : $\sqrt[3]{1,728}$

<u>Step 1:</u> $\sqrt[3]{1,728} =$	We make groups of three digits starting from the right. We have 2 groups here, so 2 digits in CR.
<u>Step 2:</u> $\sqrt[3]{1, \ 728}$ $= \quad \_\,2$	Last digit in the number is 8 So last digit in CR = 2 (as $2^3 = 8$).
<u>Step 3:</u> $\sqrt[3]{1, \ 728}$ $= 1 \quad 2$	First group is 1 $1^3 = 1$ and $2^3 = 8$ So first digit in CR = 1 Answer : $\sqrt[3]{1, \ 728} = 12$

Example 2 : $\sqrt[3]{13824}$

<u>Step 1:</u> $\sqrt[3]{13,824} =$	We make groups of three digits starting from the right. We have 2 groups here, so 2 digits in CR.

Step 2:
$3\sqrt{13,824}$

$= \_\_4$

Last digit in the number is 4
So last digit in CR = 4 (as $4^3 = 64$).

Step 3:
$3\sqrt{13,824}$

$= 2\ 4$

First group is 13
$2^3 = 8$ and $3^3 = 27$
So first digit in CR = 2
Answer : $3\sqrt{13,824} = 24$

Example 3 : $3\sqrt{175616}$

$3\sqrt{175,616}$

$=\ 5\quad 6$

(1) We have 2 groups here, so 2 digits in CR
(2) Last digit in the number is 6
     So last digit in CR = 6
(3) First group is 175
     $5^3 = 125$ and $6^3 = 216$
     So first digit in CR = 5

Answer : $3\sqrt{175,616} = \mathbf{56}$

Example 4 : $3\sqrt{681472}$

$3\sqrt{681,472}$

$=\ 8\quad 8$

(1) We have 2 groups here, so 2 digits in CR
(2) Last digit in the number is 2
     So last digit in CR = 8
(3) First group is 681
     $8^3 = 512$ and $9^3 = 729$
     So first digit in CR = 8

Answer : $3\sqrt{681,472} = \mathbf{88}$

Given below are a few more examples solving sums in straight steps :

Example 5 : $3\sqrt{17,576}$

$= 2, 6$ (as $2^3 = 8, 3^3 = 27$)

Answer : **26**

Example 6 : $3\sqrt{39,304}$

$= 3, 4$ (as $3^3 = 24, 4^3 = 64$)

Answer : **34**

Example 7 : $^3\sqrt{389,017}$

= 7 , 3 (as 7^3=343, 8^3=512)

Answer : 73

Example 8 : $^3\sqrt{1061,208}$

= 10, 2 (as 10^3=1000, 11^3=1331)

Answer : 102

Rules :

(1) Make groups of three digits in the number, starting from the right.

(2) We find out the last digit in the Cube root by observing the last digit in the number.

(3) By observing the first pair of the number we find out the first digit of the cube root.

(4) This method of finding the cube roots is possible for exact cubes with a maximum of six digits.

Exercise 16 :

Find the cube roots of the following exact cubes :

(1) $^3\sqrt{4913}$

(2) $^3\sqrt{13824}$

(3) $^3\sqrt{24389}$

(4) $^3\sqrt{46656}$

(5) $^3\sqrt{195112}$

(6) $^3\sqrt{74088}$

(7) $^3\sqrt{456533}$

(8) $^3\sqrt{531441}$

(9) $^3\sqrt{1030301}$

(10) $^3\sqrt{857375}$

Answers

Exercise 16			
(1) 17	(2) 24	(3) 29	(4) 36
(5) 58	(6) 42	(7) 77	(8) 81
(9) 101	(10) 95		

FLAME IN A TRIANGLE

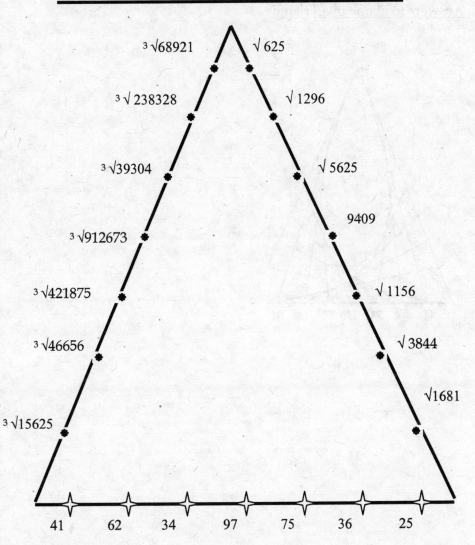

Find the Cube Roots and Square Roots
of the numbers given and join them to their
corresponding answers in the base of the triangle

Answers to Flame in a Triangle

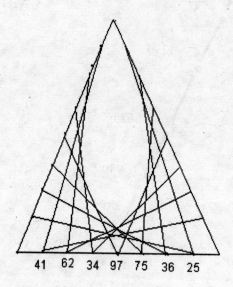

41   62   34   97   75   36   25

Chapter 17

STRAIGHT DIVISION

Till now we have studied division by numbers, which are close to the base, either more or less. But now we will learn the general method of division.

First let us take a few examples first to understand straight division by single digits.

Example 1 : 4327/3

Step 1:

```
3 | 4 3 2 7
  |   1
  └─────────
      1
```

We divide 4 by 3 , we get Quotient (Q) = 1 , Remainder (R) = 1 we place the quotient digit below and carry the remainder digit and prefix it to the second dividend digit and make it 13, our new dividend.

Step 2:

```
3 | 4 3 2 7
  |   1 1
  └─────────
      1 4
```

We now divide 13 by 3 and get 4 as quotient digit and 1 as remainder digit. Prefix the remainder to 2 and make it 12

Step 3:

```
3 | 4 3 2 7
  |   1 1 0
  └─────────
      1 4 4
```

We now divide 12 by 3 and get 4 as quotient digit and 0 as remainder digit. Prefix it to 7.

Step 4:

```
3 | 4 3 2 7
  |   1 1 0
  └─────────
    1 4 4 2 / 1
```

Now divide 7 by 3 we get 2 as quotient digit and 1 as remainder digit Since there is no more dividend digit, this is the final remainder.

Answer : 4327/3 gives Q = 1442 , R = 1

Example 2 : 6344/7

Step 1:

```
7 | 6 3 4 4
  |     0
  9
```

Here 6 cannot be divided by 7, so we divide 63 by 7 and we get,
quotient = 9, remainder = 0.
We place the quotient digit below and carry the remainder digit and prefix it to the second dividend digit and make it 04

Step 2:

```
7 | 6 3 4 4
  |   0 4
  9 0
```

We now divide 04 by 7 and get
0 as quotient digit and
4 as remainder digit.
Prefix the remainder to 4 and make it 44

Step 3:

```
7 | 6 3 4 4
  |   0 4
  9 0 6 / 2
```

We now divide 44 by 7 and get
6 as quotient digit and
2 as remainder digit.
This is the final remainder.

Answer : 6344/7 gives **Q = 906 , R = 2**

Given below are a few more examples solving sums in straight steps :

Example 3 : 893 / 4

```
4 | 8 9 3
  |   0 1
  2 2 3 / 1
```

Answer : **Q = 223 , R = 1**

Example 4 : 5032 / 8

```
8 | 5 0 3 2
  |     2 7
  6 2 9 / 0
```

Answer : **Q = 629 , R = 0**

Exercise 17.1

Divide the following :-

(1) 2330/6

(2) 4826/4

(3) 739278/8

(4) 634092/7

(5) 23143/3

(6) 3468/4

(7) 826371/7

(8) 630372/9

(9) 6029318/8

(10) 29384/5

Division by Double digit Numbers

Till now we have studied division where the divisor is a single digit but in the case of a double-digit divisor we will follow the same basic method of a single digit but with a little change. For this method we will use the

ध्वजांक

"Dhvajāṅka"

"On the top of the flag"

For this we write the first digit of the divisor as the base digit of the divisor and place all the other digits at the flag position of the first digit. The first digit alone performs division. In case of a double-digit divisor we get only one flag digit, but in the case of a triple digit divisor we get 2 flag digits and so on.

Example 1 : 756/21

Step 1:

2^1 | 7 5 / 6

Here we have a double-digit divisor so we write the first digit (2) down and place the other digit (1) as a flag of 2 (first digit) as shown.
2 is the divisor digit and 1 is the flag digit

Number of flag digits = Number of digits in the remainder side
So here we have 1 remainder digit ,
We will divide only by the base digit of the divisor.

Step 2:

2^1 | 7 5 / 6
　　| 1
　　|_____
　　3

7 divided by 2, gives
Q = 3 , R = 1
Write 3 as the first quotient digit and the remainder digit is prefixed to the next digit in the dividend to give 15 as our next dividend digit.

Step 3:

$$2^1 \mid 7\ 5\ /\ 6$$
$$1\quad 0$$
$$3\ 6$$

Multiply the quotient digit with the flag digit and subtract it from the next dividend digit.
15 – {3 (Q digit) × 1 (flag digit) }
15 – 3 = 12

Divide this 12 by 2 (divisor base digit), it gives Q = 6 and R = 0

6 is the next quotient digit and 0 is prefixed to 6 to make it 06 as the next dividend digit.

Step 4:

$$2^1 \mid 7\ 5\ /\ 6$$
$$1\quad 0$$
$$3\ 6\ /\ 0$$

Multiply the quotient digit with the flag digit and subtract it from the next dividend digit.
6 – {6 (Q digit) × 1 (flag digit) }
6 – 6 = 0
As we are in the remainder side, so we do not divide it again by the divisor digit.
We get 0 as the remainder.

Answer : 756/21 gives Q = 36 , R = 0

Example 2 : 848/31

Step 1:

$$3^1 \mid 8\ 4\ /\ 8$$

Here we have a double-digit divisor so we write the first digit (3) down and place the other digit (1) as a flag of the first digit as shown.
3 is the divisor digit and 1 is the flag digit.

Number of flag digits = Number of digits in the remainder side.
Here we have 1 remainder digit.

Step 2:

$$3^1 \mid 8\ 4\ /\ 8$$
$$2$$
$$2$$

8 divided by 3 gives
Q = 2 , R = 2
Write 2 as the first quotient digit.
The remainder digit is prefixed to the next digit of the dividend to give 24 as the next dividend.

Step 3:

$$3^1 \mid \begin{array}{c} 8\ 4\ /\ 8 \\ \underline{2\quad 1} \\ 2\ 7 \end{array}$$

Multiply the quotient digit with the flag digit and subtract it from the next dividend digit.

24 – {2 (Q digit) × 1 (flag digit) }
24 – 2 = 22

Divide this 22 by 3 (divisor base digit),
it gives Q = 7 and R = 1
7 is the next quotient digit and
1 is prefixed to 8 to make it 18 as the next dividend digit.

Step 4:

$$3^1 \mid \begin{array}{c} 8\ 4\ /\ 8 \\ \underline{2\quad 1} \\ 2\ 7\ /\ 11 \end{array}$$

Multiply the quotient digit with the flag digit and subtract it from the next dividend digit.

18 – {7 (Q digit) × 1 (flag digit) }
18 – 7 = 11

As we are in the remainder side, so we do not divide it again by the divisor digit.
We get 11 as the remainder.

Answer : 848 / 31 gives Q = 27 , R = 11

Example 3 : 387/32

Step 1:

$$3^2 \mid 3\ 8\ /\ 7$$

Here we have a double-digit divisor so

3 is the divisor digit and 2 is the flag digit
One flag digit so 1 remainder digit.

Step 2 :

$$3^2 \mid \begin{array}{c} 3\ 8\ /\ 7 \\ 0 \\ \hline 1 \end{array}$$

3 divided by 3 gives
quotient = 1 , remainder = 0

1 is the first quotient digit and
0 is prefixed to 8 to make 08 as the next dividend digit.

Step 3:

$$3^2 \mid 3\ 8\ /\ 7$$
$$ 0\ \ 0$$
$$\rule{2cm}{0.4pt}$$
$$ 1\ 2$$

Now
08 – {1 (quotient digit) × 2 (flag digit) }
8 – 2 = 6

Divide this 6 by 3 (divisor base digit)
it gives quotient = 2 and remainder = 0
2 is the next quotient digit and
0 is prefixed to 7 to make it 07 as the next
dividend

Step 4:

$$3^2 \mid 3\ 8\ /7$$
$$ 0\ \ 0$$
$$\rule{2cm}{0.4pt}$$
$$ 1\ 2\ /\ 3$$

Now
7 – {2 (quotient digit) × 2 (flag digit) }
7 – 4 = 3
We are in the remainder side, so we do not
do the division part.
We get 3 as the remainder

Answer : 387/32 gives **Q = 12 , R = 3**

Let us try doing the whole sum in a single step.

Example 4 : 86369/76

Step 1:

$$7^6 \mid 8\,6\,3\,6\,9$$

Here we have 7 as the divisor digit and 6 as
flag digit.

Step 2:

$$7^6 \mid 8\ 6\ 3\ 6\ /\ 9$$
$$ 1\ 3\ 6\ \ 6$$
$$\rule{2.5cm}{0.4pt}$$
$$ 1\ 1\ 3\ 6\ /\ \ 3\ 3$$

(1) 8/7 gives Q = 1, R = 1
(2) 16 – (1 × 6) = 10
     10/7 gives Q = 1, R = 3
(3) 33 – (1 × 6) = 27
     27/7 gives Q = 3, R = 6
(4) 66 – (3 × 6) = 48
     48/7 gives Q = 6, R = 6
(5) 69 - (6 × 6) = 33 = R

Answer : 86369/76 gives **Q = 1136, R = 33**

<u>Rules</u> :

(1) In case of two digit divisors we keep the first (tens) digit as the divisor digit and place the other (unit) as the flag digit.

(2) The number of digits in the flag will give us the number of digits in the remainder side.

(3) We divide the first dividend digit by the divisor's base digit.

(4) The quotient becomes the first quotient (Q) digit and we prefix the remainder (R) digit to the next dividend digit to give the new dividend.

(5) Then this quotient digit is multiplied by the flag digit and the result is subtracted from the new dividend digit, this is then divided by the divisor's base digit to get the next quotient digit and the remainder is again prefixed to the next dividend digit.

(6) We keep dividing till we reach the remainder side.

(7) When we are in the remainder side we will just do the subtraction and not the division part to get the remainder of the problem.

Given below are a few more examples solving sums in straight steps :

Example 5 : 685/21

$$2^1 \begin{array}{|cc/c} 6 & 8 & / 5 \\ 0 & 1 & \end{array}$$

3 2 / 13

Answer : Q = 32 , R = 13

Example 6 : 458/63

$$6^3 \begin{array}{|cc/c} 4 & 5 & / 8 \\ & 3 & \end{array}$$

7 / 17

Answer : Q = 7 , R = 17

Example 7 : 9774/78

```
7 8 | 9 7 7 / 4
     |   2 5   6
     |_____
       1 2 5 / 2 4
```

Example 8 : 59266 / 53

```
5 3 | 5 9 2 6 / 6
     |   0 1 4   3
     |_____
       1 1 1 8 / 12
```

Answer : Q = 125 , R = 24   Answer : Q = 1118 , R = 12

Exercise 17 . 2 :

Attempt the following divisions in straight steps :-

(1) 247/21   (6) 4841/43

(2) 441/33   (7) 52079/25

(3) 4911/32   (8) 81800/52

(4) 5065/22   (9) 97586/66

(5) 2924/72   (10) 96001/79

Special Cases with Alternate Remainder

Till now we have studied simple examples for straight division. Let us now take a few difficult examples where we might get a negative number after subtracting the product of the quotient and flag digit from the dividend. In such cases we need to change our quotient digit in order to get a bigger remainder and thus a bigger digit as our next dividend. And so we can get a positive number after the subtraction and in such cases we use

संकलन व्यवकलनाभ्यां

"Saṅkalana Vyavakalanābhyām"

"By addition and by subtraction"

In such cases we will be using alternate remainders. Alternate remainder means that when a number is divided by another number and we get a certain remainder, but if we reduce our quotient by one digit, our remainder will increase, and this way we can change our remainder by changing the quotient. If we reduce the quotient by two digits then the remainder will increase further. As our sutra says we subtract 1 from our quotient and add 1 divisor to the remainder. This is called alternate remainder.

Let us take an example to understand it.

7/2 gives Q = 3, R = 1, it can also give
          Q = 2, R = 3,
          Q = 1, R = 5,
          Q = 0, R = 7

We will use this alternate remainder method in our division without which learning division is not complete.

Example 1 : 5614/21

Step 1:   Here we have 2 as the divisor base digit and 1 as the flag digit

$$2^1 \, | \, 5\,6\,1\,/\,4$$

Step 2:

2^1 | 5 6 1 / 4
        | 1
      2

5 divided by 2 gives

Q = 2 , R = 1

Next dividend digit becomes 16

Step 3:

2^1 | 5 6 1 / 4
        | 1 2
      2 6

Now 16 – {2 (Q digit) × 1 (flag digit)} = 14
14/2 gives Q = 7 , R = 0

01 becomes our next quotient digit.
Now 01 – {7 (Q digit) x 1(flag digit)} gives
negative number. So we will change our
dividend of this step by taking alternate
remainder.

14/2 also gives Q = 6 and R = 2
this gives the next dividend digit as 21

and 21 – {6 (Q digit) x 1(flag digit)} gives
positive number.

*In case we still get a negative number we will take
the next alternate remainder. We keep checking
this till we get a positive answer.*

Step 4:

2^1 | 5 6 1 /4
        | 1 2 1
      2 6 7

21 – (6 × 1) = 15
Now since 15 is positive we need not change
the quotient and the dividend digit of the
previous step.

15 divided by 2 gives
Q = 7 , R = 1
Next dividend digit is 14

Step 5:

2^1 | 5 6 1 /4
        | 1 2 1
      2 6 7 / 7

Since we are in the remainder side, so we will
only subtract

14 – (7 × 1) = 7
R = 7

Answer : 5614/21 gives Q = 267, R = 7

Example 2 : 6712/53

Step 1

5^3 | 6 7 1 / 2

Here we have 5 as the divisor base digit and 3 as the flag digit.

(1) 6/5 gives Q = 1 , R = 1

Step 2:

5^3 | 6 7 1 / 2
      | 1 4   5

    1 2 6 / 3 4

(2) 17 – (1 × 3) = 14
    14/5 gives Q = 2 , R = 4

(3) 41 – (2 × 3) = 35
    35/5 gives Q = 7 , R = 0

The next dividend 02 will give a negative answer when 7 (Q digit) x 3 (flag digit) is subtracted from it.

So we take alternate remainder here, 35/5 also gives Q = 6 and R = 5.

Now 52 is the next dividend.
And 52 - (6 x 3) gives a positive number.

(4) 52 – (6 × 3) = 34
    R = 34

Answer : 6712/53 gives Q = **126** , R = **34**

Example 3 : 1864/29

Step 1:

2^9 | 1 8 6 / 4

Here we have 2 as the divisor base digit and 9 as the flag digit.

<u>Step 2 :</u>   18/2 gives Q = 9 , R = 0

$$2^9 \overline{\left| \begin{array}{cc} 1\ 8\ 6\ /\ 4 \\ 6\ 4 \end{array} \right.}$$

6

The next dividend 06 will give a negative answer when 9 (Q digit) x 9 (flag digit) = 81 is subtracted from it, so we need alternate remainder here.

- 18/2 also gives Q = 8 , R = 2,
  (26 – (8 × 9) = negative)
- 18/2 also gives Q = 7 , R = 4,
  (46 – (7 × 9) = negative)
- 18/2 also gives Q = 6 , R = 6,
  (66 – (6 × 9) = positive)

Q = 6 and the next dividend digit = 66

<u>Step 3:</u>   66 – (6 × 9) = 12
                                 12/2 gives Q = 6 , R = 0

$$2^9 \overline{\left| \begin{array}{cc} 1\ 8\ 6\ /\ 4 \\ 6\ 4 \end{array} \right.}$$

6 4 / 8

The next dividend 04 will give a negative answer when (6 x 9) = 81 is subtracted from it, so we need alternate remainder here.

- 12/2 gives Q = 5 , R = 2,
  (24 – (5 × 9) = negative)
- 12/2 gives Q = 4 , R = 4,
  (44 – (4 × 9) = positive)
  Q = 4 and the next dividend digit = 44

44 – (4 × 9) = 8
R = 8

Answer : 1864 / 29 gives **Q = 64 , R = 8**

Example 4 : 4118/27

Step 1:

$$2^7 \quad \begin{array}{|l} 4\ 1\ 1\ /\ 8 \\ 2\ 4\ \ 2 \end{array}$$

$$1\ 5\ 2\ /\ 1\ 4$$

(1) 2 as the base digit, 7 as the flag digit, 8 in the remainder side.
(2) 4/2 gives Q = 2, R = 0, alternately we take Q = 1 , R = 2
(3) 21 – (1 × 7) = 14
    14/2 gives Q = 7, R = 0, alternately we take Q = 5 and R = 4
(4) 41 – (5 × 7) = 6
    6/2 gives Q = 2 , R = 2
(5) 28 – (2 × 7) = 14
    Final R = 14

Answer : 4118 / 27 gives **Q = 152, R = 14**

Rules:

(1) We take alternate remainder when we get a negative number on subtracting the product of the quotient digit and flag digit from the dividend of the next step.

(2) We keep changing our remainder till it leads to a positive number in the subtraction of the next step.

Given below are a few more examples solving sums in straight steps :

Example 5 : 3412/24   **Example 6 :** 6712/56

$$2^4 \quad \begin{array}{|l} 3\ 4\ 1\ /\ 2 \\ 1\ 2\ \ \ 1 \end{array} \qquad 5^6 \quad \begin{array}{|l} 6\ 7\ 1\ /\ 2 \\ 1\ 6\ \ 10 \end{array}$$

$$1\ 4\ 2\ /\ 4 \qquad\qquad\qquad 1\ 1\ 9\ /\ 48$$

Answer : **Q = 142, R = 4** .   Answer : **Q = 119, R = 48**

Example 7 : 1408/23

$$2^3 \begin{array}{|cc} 1 4 0 / 8 \\ 2 \quad 0 \end{array}$$

$$6 \ 1 / 5$$

Answer : Q = 61, R =5

Example 8 : 16304/63

$$6^3 \begin{array}{|cc} 1 6 3 0 / 4 \\ 4 \ 7 \quad 7 \end{array}$$

$$2 5 8 / 5 0$$

Answer : Q = 258, R = 50

Exercise 17. 3 :

Attempt the following divisions :-

(1) 4622/28

(2) 7858/33

(3) 8218/33

(4) 7818/22

(5) 5711/54

(6) 1876/32

(7) 15707/66

(8) 4143/18

(9) 25791/64

(10) 59754/23

Answers

Exercise 17.1
(1) Q-388, R-2	(2) Q-1206, R-2	(3) Q-92409, R-6	(4) Q-90584, R -4
(5) Q-7714, R-1	(6) Q-867, R-0	(7) Q-118053,R-0	(8) Q-70041, R-3
(9) Q-753664,R-6	(10) Q-5876, R-4		

Exercise 17.2
(1) Q-11, R-16	(2) Q-13, R-12	(3) Q-153, R-15	(4) Q-230, R-5
(5) Q-40, R-44	(6) Q-112, R-25	(7) Q-2083, R-4	(8) Q-1573, R-4
(9) Q-1478, R-38	(10) Q-1215, R-16		

Exercise 17.3
(1) Q-165, R-2	(2) Q-238, R-4	(3) Q-249, R-1	(4) Q-355, R-8
(5) Q-105, R-41	(6) Q-58, R-20	(7) Q-237, R-65	(8) Q-230, R-3
(9) Q-402, R-63	(10) Q-2598, R-0		

Chapter 18

SQUARE ROOTS II

In Chapter 15 we have studied how to find square roots of exact squares which have a maximum of 4 -5 digits. In this chapter we will study how to find square roots of any random number up to as many decimal places as we want.

We have already learned that to find the square root of any number
  (1) We make pairs of 2 digits starting from the right.
  (2) If the number has n digits then the square root will have either $n/2$ if n is even and $(n +1) / 2$ digits if n is odd.
  (3) The first digit of the square root can be found by observing the first pair of the number.

The method used is the same as that of straight division studied in the previous chapter but with a slight difference
  (1) Instead of the base divisor digit, the divisor will be the double the first digit of the square root.
  (2) Instead of subtracting the product of the quotient digit and flag digit (here there is no flag digit), the duplex of the second digit onwards of the quotient is subtracted from the new divisor in each step.
  (3) The first digit of the square root can be found by observing the first pair. It will always be a single digit, so our divisor will never be more than 18 (as a single digit can never be more than 9 and the divisor is double the first digit of the square root).

Let us understand it with a few examples:

Example 1 : $\sqrt{3249}$

<u>Step 1:</u>

32 : 4 9

The number has 4 digits so the square root will have 4/2 = 2 digits.

We make pairs of 2 digits starting from the right. We have 32 in the first pair on the left so we place a colon after 32 as shown.

Step 2:

```
      | 32 : 4 9
10:   |      7
      |_____
  5 :
```

We know that $5^2 = 25$ and $6^2 = 36$
so we get 5 as the first digit of the quotient
and remainder = 7.
We place a colon after 5 in the answer.
Double of 5 = 10, this is our divisor.
Next dividend = 74 (The remainder is
prefixed to the next dividend digit).

Step 3:

```
      | 32 : 4  9
10:   |      7  4
      |_____
  5 : 7.
```

No subtraction is performed in this step we
will do only division
Divide 74 by 10 (divisor)
74/10 gives
Q = 7, R = 4
7 is the next quotient digit and
49 is the next dividend (4 is prefixed to 9).

We needed 2 digits in the square root (step 1),
which we already have but we have further
digits in the dividend to be divided so we
carry on our division. We place the decimal
after 7 and the other quotient digits will be
placed after the decimal.

Step 4:

```
      | 32 : 4  9
10:   |      7  4
      |_____
  5 : 7. 0
```

Now from the new dividend 49 the duplex of
the second quotient digit 7 (first digit after
the colon) is subtracted to get the new
dividend, which is divided by 10.
    49 – D (7)
  $= 49 – 7^2 = 0$
and 0/10 gives Q = 0, R = 0

Since we already have 2 digits in the square
root, this quotient digit (0) is placed after a
decimal.
Our division is complete as we do not have
any further digit in the dividend to be
divided and we are getting 0 as our quotient
digits after the decimal, and zero is the
remainder, so our square is a complete
square.
Answer : $\sqrt{3249} = 57$

Example 2 : $\sqrt{119716}$

Step 1:

11 : 9 7 1 6

The number has 6 digits so the square root will have 6/2 = 3 digits.
We make pairs of 2 digits starting from the right.
We have 11 in the first pair on the left so we place a colon after 11 as shown.

Step 2:

$$\begin{array}{c|c} & 11 : \ 9716 \\ 6: & \quad\ 2 \\ \hline & 3 : \end{array}$$

We know that $3^2 = 9$ and $4^2 = 16$
so we get 3 as the first digit of the quotient and remainder = 2,
We place a colon after 3 in the answer.
Double of 3 = 6 as our divisor
Next dividend = 29 (2 is prefixed to 9)

Step 3:

$$\begin{array}{c|c} & 11 : \ 9\ 716 \\ 6: & \quad\ 2\ 5 \\ \hline & 3 : 4 \end{array}$$

No subtraction is performed in this step, we will perform only division.

Divide 29 by 6 (divisor)
29/6 gives
Q = 4 , R = 5
4 is the next quotient digit
Next dividend = 57 (5 is prefixed to 7)

Step 4:

$$\begin{array}{c|c} & 11 : \ 9\ 7\ 1\ 6 \\ 6: & \quad\ 2\ 5\ 5 \\ \hline & 3 : 4\ 6 \end{array}$$

Now from the new dividend 57 the duplex of the second quotient digit 4 (first digit after the colon) is subtracted to get the new dividend, which is divided by 6.
57 – D (4)
= 57 – 4^2 = 41
Divide 41 by 6 (divisor)
41/6 give
Q = 6 , R = 5
6 is the next quotient digit and
Next dividend = 51 (5 is prefixed to 1)

We needed 3 digits in our square root (step 1), which we already have but as we still have further digits in the dividend to be divided so we carry on our division and place all other digits after a decimal.

Step 5:

$$11: \ 9\,7\,1\ 6$$
$$6: \quad\quad 2\,5\,5\ 3$$
$$\overline{}$$
$$3: \ 4\,6\,.0$$

Now from the new dividend 51, the duplex of the second and third digits 46 (both the digits after the colon) is subtracted to get the new dividend, which is divided by 6.

$$51 - D\,(46)$$
$$= 51 - 2\,(4 \times 6)$$
$$= 51 - 48 \ = 3$$

and 3/6 gives

$$Q = 0\,, R = 3$$

0 is the next quotient digit and

Next dividend = 36 (3 is prefixed to 6)

Since we already have 3 digits in the square root this quotient digit (0) is placed after a decimal.

Step 6:

$$11: \ 9\ 7\ 1\ 6$$
$$6: \quad\quad 2\ 5\ 53$$
$$\overline{}$$
$$3: \ 4\,6\,.0\,0$$

Now from the new dividend 36 the duplex of all digits after the colon 460 is subtracted to get the new dividend, which is divided by 6.

$$36 - D\,(460)$$
$$= 36 - (\,2(4 \times 0) + 6^2\,)$$
$$= 36 - 36 \ = 0$$

and 0/6 gives

$$Q = 0\,, R = 0$$

0 is the next quotient digit.
Our division is complete as we do not have any further digit in the dividend to be divided. Since we are getting zeros as the quotient digits after the decimal, and zero as the remainder, so our square is a complete square.

Answer : $\sqrt{119716} = 346.00 =$ **346**

Example 3 : $\sqrt{53163214}$

Step 1:

$$53 : 16\ 32\ 14$$

The number has 8 digits so the square root will have 8/2 = 4 digits

We make pairs of 2 digits starting from the right.

We have 53 in the first pair on the left so we place a colon after 53 as shown.

Step 2

$$\begin{array}{c|l} & 53:\ 1\ 6\ 3\ 2\ 1\ 4 \\ 14: & \underline{\quad 4 \quad} \\ & 7: \end{array}$$

We know that $7^2 = 49$ and $8^2 = 64$ so we get 7 as the first digit of the quotient and remainder = 4

We place a colon after 7 in the answer.

Divisor = 14

Next dividend = 41

Step 3:

$$\begin{array}{c|l} & 53:\ 1\ \ 6\ 3\ 2\ 1\ 4 \\ 14 & \underline{\quad 4\ \ 13 \quad} \\ & 7:2 \end{array}$$

No subtraction is performed in this step

41/14 gives
Q = 2 , R = 13

Next dividend = 136

Step 4:

$$\begin{array}{c|l} & 53:1\ 6\ \ 3\ 2\ 1\ 4 \\ 14 & \underline{\quad 4\ 13\ \ 6 \quad} \\ & 7:2\ 9 \end{array}$$

136 – D (2)
= 136 – 2^2 = 132

and 132/14 gives
Q = 9 , R = 6

9 is the next quotient digit and next dividend = 63

Step 5:

$$\begin{array}{c|l} & 53:1\ 6\ 3\ \ 2\ 1\ 4 \\ 14 & \underline{\quad 4\ 13\ 6\ \ 13 \quad} \\ & 7:2\ 9\ 1 \end{array}$$

63 – D (29)
= 63 – 2 (2 x 9) = 27

and 27/14 gives
Q = 1 , R = 13

1 is the next quotient digit and next dividend = 132

Step 6:

$$\begin{array}{c|l} & 53:1\ 6\ 3\ \ 2\ 1\ 4 \\ 14 & \underline{\quad 4\ 13\ 6\ 13\ 5 \quad} \\ & 7:2\ 9\ 1.\ 3 \end{array}$$

132 – D (291)
= 132 – (2(2 x 1) + 9^2)
= 47

and 47/14 gives
Q = 3 , R = 5

3 is the next quotient digit and
next dividend = 51
Since we should have 4 digits in the square
root this quotient digit (3) is placed after a
decimal.

Step 7:

$$53 : 1\ 63\ 21\ 4$$
$$14 \quad 4\ 13\ 6\ 13\ 5\ 7$$

$$7 : 2\ 9\ 1.\ 3\ 1$$

51 – D (2913)
= 51 – (2(2 × 3) + 2(9 × 1))
= 51 – 30 = 21

and 21/14 gives
Q = 1 , R = 7

1 is the next quotient digit and
next dividend = 74

Step 8:

$$53 : 1\ 6\ 3\ 21\ 4$$
$$14: \quad 4\ 13\ 6\ 13\ 5\ 7$$

$$7 : 2\ 9\ 1.\ 3\ 1\ 1$$

74 – D (29131)
= 74 – (2(2×1) + 2(9×3) + 1²)
= 15

and 15/14 gives
Q= 1 , R = 1
1 is the next quotient digit.

This is not a complete square as we have
performed division on all the digits of the
dividend and we are still getting further
digits for the quotient.
We have found our answer up to 3 decimal
places and we can continue in the same way
to find more decimal places by adding zeros
to our dividend and further performing the
division on them.

Answer : √ 53163214 = 7291. 311

Example 4 : √ 16384

Step 1:

$$1 : 63\ 84$$

The number has 5 digits so the square root
will have (5 +1)/2 = 3 digits

We make pairs of 2 digits starting from the right.
We have 1 in the first pair on the left so we place a colon after 1 as shown

Step 2:

$$\begin{array}{l|l}
 & 1:6\ 3\quad 8\ 4 \\
2: & \quad 0 \\
\hline
 & 1: \\
\end{array}$$

We know that $1^2 = 1$ and $2^2 = 4$
so we get 1 as the first digit of the quotient and remainder = 0,

We place a colon after 1 in the answer.
Divisor = 2
Next dividend = 06

Step 3:

$$\begin{array}{l|l}
 & 1:6\ 3\quad 8\ 4 \\
2: & \quad 0\ 2 \\
\hline
 & 1:2 \\
\end{array}$$

No subtraction is performed in this step
06/2 gives
Q = 3 , R = 0

03 – D (3) gives negative number, so we use alternate remainder.

06/2 also gives Q = 2 , R= 2
2 is the next quotient digit and
next dividend = 23

Step 4:

$$\begin{array}{l|l}
 & 1:6\ 3\ 8\ 4 \\
2: & \quad 0\ 2\ 3 \\
\hline
 & 1:2\ 8 \\
\end{array}$$

  23 – D (2)
= 23 – 2^2 = 19
and 19/2 gives Q = 9 , R = 1

18 – D(29) gives negative number, so we use alternate remainder.
19/2 also gives Q = 8 , R= 3

8 is the next quotient digit and
next dividend = 38

Step 5:

$$\begin{array}{l|l}
 & 1:6\ 3\quad 8\ 4 \\
2: & \quad 0\ 2\ 3\ 6 \\
\hline
 & 1:2\ 8\ .0 \\
\end{array}$$

  38 – D (28)
= 38 – 2 (2 × 8) = 6
and 6/2 gives Q = 3 , R = 0

04 - D(283) gives negative number, so we use alternate remainder

6/2 also gives
Q =2, R = 2 ,
(24 – D(282) gives negative number)
Q =1, R = 4,
(44 – D(281) gives negative number)
Q = 0 , R= 6 ,
(64 - D(280) gives a positive number)

0 is the next quotient digit and
next dividend = 64

Since we should have 3 digits in the square root this quotient digit 0 is placed after a decimal.

Step 6:

$$\begin{array}{c|l} & 1:6\ 3\ 8\ 4 \\ 2: & 0\ 2\ 3\ 6 \\ \hline & 1:2\ 8\ .0 \end{array}$$

$64 - D (280)$
$= 64 - (2 (2 \times 0) + 8^2) = 0$

and 0/2 give
Q = 0 , R = 0

Our division is complete.

Answer : $\sqrt{16384} = \mathbf{128}$

Rules :

(1) We make pairs of 2 digits starting from the right.

(2) If the number has n digits then the square root will have $n/2$ digits if n is even, and $(n +1)/2$ digits if n is odd.

(3) The first digit of the square root can be found out by observing the first pair of digits of the number. Place a colon after the first digit of the answer.

(4) The divisor is double the first digit of the square root.

(5) The duplex of the second digit onwards is subtracted from the new divisor in each step.

(6) If we are getting a negative number after the subtraction of the duplex from the dividend then we use an alternate remainder to get the next dividend.

(7) Our division is complete, if we get a zero for both the quotient and the remainder, otherwise we have an irrational answer.

We give below a few more examples solving sums in straight steps

Example 5 : $\sqrt{1389}$

	13 : 8 9 0 0
6	4 6 8 16

3 : 7 . 2 6

Answer : **37.26**

Example 6 : $\sqrt{724201}$

	72 : 4 2 0 1
	8 4 1 0

8 : 51 . 0 0

Answer : **851**

Example 7 : $\sqrt{25745476}$

	25 : 7 4 5 4 7 6
10	0 7 4 5 5 1

5 : 0 7 4 . 0 0 0

Answer : **5074**

Example 8 : $\sqrt{28}$

	28 : 0 0 0 0 0
10	3 10 6 14 5

5 : . 2 9 1 5

Answer : **5 . 2915**

Exercise 18 :

Find the square roots of the following numbers :-

(1) √ 4096   (6) √ 732108

(2) √ 552049   (7) √ 18134512

(3) √ 20828   (8) √ 1387

(4) √ 140213   (9) √ 42406144

(5) √ 329752   (10) √ 4782969

Answers

Exercise 18			
(1) 64	(2) 743	(3) 144.319	(4) 374.45
(5) 574.24	(6) 855.633	(7) 4258.463	(8) 37.242
(9) 6512	(10) 2187		

List of all Sutras used in the text with their Applications

Glossary

Alternate Remainder: When a number is divided by another number and we get a certain remainder, and then if we reduce our quotient by one digit, our remainder will increase. In this way we can change our remainder by changing the quotient. This is called Alternate Remainder. (eg. 6/3 gives Q - 2,R- 0, alternately Q -1,R-3)

Close to base: The numbers which have a small complement –slightly more or less than the base. (eg. 87, 76, 104, 112)

Coefficient: Coefficient is a constant multiplicative factor of a certain object. (eg 3x, 5y, here 3 and 5 are the coefficients of x and y respectively)

Complement: Any number when subtracted from its nearest base will give its complement. (eg. complement of 87 is 13)

Composite numbers: A composite number is a positive integer which has a positive divisor other than one or itself. (eg. 24, 36)

Geometric ratio: The ratio of two consecutive terms in a geometric progression. (eg. 2, 4, 8 are in geometric ratio)

L.H.S.: Left Hand Side.
R.H.S.: Right Hand Side.

Natural numbers: A positive whole numbers.(eg.1,2,3,4…)

Place value: The value given to a digit by virtue of its location in a numeral. (Place value of 4 in 645 if 40)

Prime numbers: A prime number (or a prime) is a natural number that has exactly two (distinct) natural number divisors, which are 1 and the prime number itself. (eg. 13,17,47)

Proportional: Two quantities are called proportional if they vary in such a way that one of the quantities is a constant multiple of the other, or equivalently if they have a constant ratio. (eg. The circumference of the circle is proportional to its diameter.)

Variable: A variable is a symbol denoting a quantity or symbolic representation. In mathematics, a variable often represents an *unknown* quantity that has the potential to change. (eg. 3x,5y, here x and y are the variables)

INDEX

Activities are marked with a *